Also available from Routledge

Facilities Management: Theory and Practice
Keith Alexander

Pb: 0-419-20580-2

Adapting Buildings for Changing Uses
David Kincaid

Pb: 0-419-23570-1

The Dynamics of Property Location
Russell Schiller

Hb: 0-415-24645-8

Property Development, 4th edition
D. Cadman and R. Topping

Pb: 0-419-20240-4

Information and ordering details

For price availability and ordering visit our website **www.routledge.com**
Alternatively our books are available from all good bookshops.

Facilities Management
Innovation and Performance

Edited by

Keith Alexander, Brian Atkin, Jan Bröchner and Tore I. Haugen

Routledge
Taylor & Francis Group

LONDON AND NEW YORK

First published 2004
by Spon Press

This edition published 2012 by Routledge
2 Park Square, Milton Park, Abingdon, Oxon, OX14 4RN
711 Third Avenue, New York, NY 10017

Routledge is an imprint of the Taylor & Francis Group, an informa business

© 2004 Taylor & Francis

Publisher's Note
This book is produced from camera-ready copy supplied from the editors

Every effort has been made to ensure that the advice and information in
this book is true and accurate at the time of going to press. However,
neither the publisher nor the authors can accept any legal responsibility or
liability for any errors or omissions that may be made. In the case of drug
administration, any medical procedure or the use of technical equipment
mentioned within the book, you are strongly advised to consult the
manufacturer's guidelines.

British Library Cataloguing in Publication Data
A catalogue record for this book is available from the British Library

Library of Congress Cataloging in Publication Data
A catalog record for this book has been requested

ISBN 978-0-415-32146-4

Contents

Contributors

Professor Keith Alexander

Keith Alexander is Professor of Facilities Management and Director of the Centre for Facilities Management (CFM) at the University of Salford, Greater Manchester, UK. Past Chairman of the European Facility Management Network (EuroFM), Keith has lead important national, European and international initiatives in education, research and practice. He created one of the first Masters programmes in Facilities Management in the world in 1986, was a founder and the first elected Chairman of the European Facility Management Network (EuroFM) and created the International Faculty of Facilities Management (IFFM). He has also contributed to the development of international standards for the recognition of FM courses and the accreditation of FM professionals (IFMA).

Professor Jan Bröchner

Jan Bröchner is Professor of Organisation of Construction, specialising in Facilities Management, at Chalmers University of Technology in Göteborg, Sweden. He is currently the regional editor for Western Europe of Facilities, and has been a board member of the EuroFM Network.

Professor Brian Atkin

Brian Atkin PhD, MPhil, BSc, FRICS, FCIOB is Programme Director for the Swedish national construction R&D programme, *Competitive Building*, a position he holds through Lund Institute of Technology, part of Lund University. He holds or has held professorial appointments at the University of Reading in the UK, VTT (Technical Research Centre of Finland), Helsinki University of Technology, Royal Institute of Technology in Stockholm, Chalmers University of Technology in Gothenburg and the University of Hong Kong. He is a Director of Atkin Research & Development Limited, a specialist consultancy.

Professor Tore I. Haugen

Tore I. Haugen is Professor of Architectural Management at the Norwegian University of Science and Technology NTNU. He has a dr.ing degree in Facilities Management, and has various experiences in research and development work related both to Architectural Management and Facilities Management. He has been chairman of EuroFM Research Forum, a chairman of ISO TC59/SC13 Information Classification and is active both in different national and international research networks. He is currently in charge of the research and educational program Metamorfose Real Estate and Facilities Management at NTNU.

Dilanthi Amaratunga

Dilanthi Amaratunga is the Director of Postgraduate Research Studies at Research Institute for the Built and Human Environment at the University of Salford. She leads a CIB TG 53 on Postgraduate Research Training in Building and Construction. Dilanthi has achieved widespread recognition of her work in facilities management and process improvement through an extensive number of published international refereed journal papers and by means of presentations made at major conferences in both UK and overseas.

Kirsten Arge

Kirsten Arge (kirsten.arge@byggforsk.no) is an architect by education and is working as a senior researcher at the Norwegian Building Research Institute in Oslo. Her research is centred on property development, management and use.

David Baldry

David Baldry is a Senior Lecturer in the School of Construction and Property Management, University of Salford, UK. He is a Chartered Surveyor and a member of the Research Information, and Knowledge Committee of the British Institute of Facilities Management.

Paul Dettwiler

Paul Dettwiler, Licentiate of Engineering, MBA, M.Sc.Arch., SAR/MSA at Department of Service Management at Chalmers University of Technology in Göteborg, Sweden.

Michael Fenker

Michael Fenker, Arch., Ph. D., is a researcher at the Laboratory Espaces – Travail, School of Architecture Paris-La-Villette. His research focuses on the relationships between business strategies, contextual factors and the design and use of workplaces.

Reidar Gjersvik

Reidar Gjersvik (Reidar.Gjersvik@sintef.no) is a Senior Researcher at SINTEF Industrial Management, Knowledge and Strategy.

Bob Grimshaw

Bob Grimshaw is Professor of Facilities Innovation at UWE Bristol. He is the Director of the Construction and Property Research Centre and manages the Facilities Innovation Programme, a collaborative research venture with Johnson Control IFM. His research is focused on the relationship between physical space and social interaction, especially in the workplace.

Linariza Haron

Linariza Haron, lriza@usm.my, is a lecturer in the Construction and Project Management programmes at the School of Housing, Building and Planning, Universiti Sains Malaysia, Penang, Malaysia. Her teaching and research interests are in property & facility management and in new ways of working.

John Hinks

John Hinks is *Innovation Manager* for Facilities & Logistics with Royal Bank of Scotland. Previously Professor of Facilities Management with Glasgow Caledonian University, John has been closely involved with the consolidation of the FM profession in the UK, and with the development of the global FM research community.

John Hudson

John Hudson (j.hudson@salford.ac.uk) is a lecturer in the School of Construction and Property Management at the University of Salford, UK. His recent work has focused on sustainable urban development in relation to facilities management and building use.

Siri H. Blakstad

Siri H. Blakstad (Siri.H.Blakstad@sintef.no) holds a PhD in adaptability and office design, and is now Research Director at SINTEF Civil and Environmental Engineering, Department of Architecture and Building Technology. Her current research aims at using knowledge from both organisational development and workplace design to gain new insight into how organisational and individual performance are affected by design and by changes in physical environment.

Margaret-Mary Nelson

Margaret-Mary Nelson is a Research Fellow in Facilities Management at the Centre for Facilities Management, University of Salford. Originally from a property management background, the mum of two is currently undertaking PhD studies in the field of Supply Chain Management in Facilities Management.

Preface

The discipline of facilities management challenges many of the norms that we associate with optimisation of the performance of a business. Successful facilities managers need a range of skills in order to devise appropriate innovative strategies for the future of the organisations in which they work. Two key concepts that have come to the fore in recent years are innovation and performance. While the initial exuberance related to Offices of the Future has been replaced with a sober appraisal of what modern information and communication technologies require from productive workplaces, there remains a strong emphasis on innovation, tempered with an increasing concern for strong performance, whether measurable or not.

In many ways, the present volume develops issues that were foreshadowed in my 'Facilities Management: Theory and Practice', first published in 1996. But some are clearly new, not least since there is a heavier emphasis on knowledge in the workplace. For the present volume, the contributions have been organised according to four strategic themes: (i) Organisational Change and Learning, (ii) Innovation and the Innovative Workplace, (iii) Performance, and (iv) Towards Knowledge Workplaces. Together, they provide a broad overview of where the field of facilities management is moving today.

Keith Alexander
University of Salford

CHAPTER 1

Introduction

Keith Alexander, Brian Atkin, Jan Bröchner and Tore I. Haugen

1.1 THE SALFORD SYMPOSIUM

The agenda for the First International Research Symposium on Facilities Management, organised by the University of Salford on behalf of the European Facility Management network, EuroFM, and held in Salford in April 2002, sought to encourage the sharing of theoretical and practical knowledge amongst researchers, and to focus on the workplace as a broad theme. The aim of the Symposium was to strengthen the theoretical foundations, to advance knowledge and to promote research into facilities management.

The event brought together leading researchers in facilities management from around the world and provided an opportunity for discussing research proposals, methods and techniques, for debating theoretical perspectives, and for reporting research in progress, research projects and their findings. Ongoing postgraduate research was also presented at the International Postgraduate Conference, University of Salford, immediately following the Symposium.

The Symposium provided an opportunity to explore Facilities Management as a strategic discipline and to develop 'the workplace' as a concept from a research perspective, as well as considering perspectives on the workplace of the future in order to identify the need for innovation and performance. In line with the objectives of the EuroFM Network, the Symposium also sought to identify links from research into practice and education.

At the outset of the Symposium, key concepts and definitions were clarified in order to provide a framework for discussion in an interactive event and to address the nature of workplace knowledge. This introductory chapter reflects the keynote presentations delivered at the Symposium by Peter Barrett and Keith Alexander, both at the University of Salford, and by Tore I. Haugen of the Norwegian University of Science and Technology (Trondheim). The chapter also sets out the structure underlying the programme for the event and introduces the four main parts of the present volume.

1.1.1 The EuroFM research agenda

The Salford symposium followed on a workshop held in Copenhagen in June 2001, which focused on flexibility in relation to workspace and office buildings. The May 2003 Symposium in Rotterdam has presented another opportunity for researchers and practitioners to take part in the further development of the facilities management research agenda in order to create scientifically based education and professionalism in Europe.

The EuroFM Network promotes interaction between research, education and practice in forming and developing the knowledge base of facilities management in Europe. The EuroFM network has been active over a period of fifteen years, and in 1990, it held the first European conference in facilities management in Glasgow, with the aim of developing facility management research and education into a more mature activity.

As part of the work with the EuroFM strategic plan, a workshop was held in Vienna in August 1999. The overall theme for the Vienna workshop was: *Where do we want to be in five years' time?* A Research Network Group was formed within EuroFM to enable it to be the leading international think tank and international knowledge base on facilities management in Europe. The major role of the research group is to establish an active research network in Europe reflecting an integrated approach to facilities management research, practice and education. The group is responsible for arranging a research forum as part of the annual EuroFM conference, and for being a generator of European research projects. This means being active in formulating the future agenda for facilities management research in the European Union, and for securing necessary documentation and information exchange between scientific research and development in practice.

The majority of the members of the Research Network Group are drawn from research organisations and educational institutions, but input from, and close co-operation with, practice are vital for adequate and high quality research and development work. Together with the Educational Group, the Research Group creates a network for collaboration between graduate facilities management programmes in European countries, and forms a link between the EuroFM members that are national organisations and the Research Group Projects.

EuroFM research activities have included collaborative research with FM Centres in the UK, the Netherlands and Sweden, and with other research organisations and educational institutions working in the area. European funded research like the *Office* and the *Workspace* projects (see Chapter 6) and the EuroFM *Benchmark* project are results of this network activity.

Since the first conference in Glasgow in 1990, we have seen a shift from a major focus on buildings and technology to a stronger focus on management of facilities in a dynamic, life long perspective. New research programmes and projects in several European countries reflect this change.

Facilities management research and development projects in EuroFM focus typically on the integration of all support activities, in time, with the core business. Facilities management research is about organisational processes (maintenance and operation of the facilities management services) and about the development and change of facilities as an integrated service. EuroFM has identified a number of relevant research topics in facilities management as its basis for further work:

- facilities management strategy (i.e. mission and business),
- facilities management structure (i.e. organisation and process),
- workspace design and management,
- facility concepts for accommodation,
- KPI/benchmarking,

- asset management (at the corporate level),
- operation and maintenance management,
- service management and quality,
- outsourcing—contracting out,
- e-commerce in facilities management,
- marketing of facilities management,
- total quality management (TQM),
- information and communication technology,
- life cycle value/profit,
- environmental strategies, and
- post occupancy evaluations (POE).

Over the years, there have been several EuroFM publication channels for research: the International Journal of Facility Management, the EuroFM practice books 1996 –1999 and Symposium Proceedings, as well as EuroFM reports, such as the 12-volume set of reports from the Workspace Thematic Networks Project under Brite-EuRam III (EuroFM, 2001). Beginning in 2001, the major information channel has been the EuroFM web site http://www.eurofm.org. Here, members of the EuroFM member organisations can access recent information from the three different network groups: Practice, Education and Research.

1.2 KEY CONCEPTS

1.2.1 Strategically integrated facilities management

In his keynote introduction to the 2002 symposium, Peter Barrett drew upon basic concepts from earlier research (Barrett, 1995; Barrett and Baldry, 2003) to provide a framework for discussion. He defined facilities management as 'a strategically integrated approach to maintaining, improving and adapting the buildings and supporting services of an organisation in order to create an environment that strongly supports the primary objectives of that organisation'.

Barrett argued that facilities management needs to embrace strategically orientated, continuous improvement and proposed key features of effective facilities management systems. Drawing upon systems theory, he suggested that such systems should be 'objective-nested, client/stakeholder orientated, minimalistic/holistic and evolutionary'.

He introduced a generic model for facilities management systems based on a combination of systems theory and information processing perspectives, and illustrating the range of continuing interactions, which are involved in facilities management. This generic model shows how an ideal facilities department would interact with the core business and the external environment. The model differentiates between strategic and operational facilities management, highlighting the need to consider both current and future situations. Barrett underlined in a corporate setting the central integrating role of facilities management functions, which guide and inform the links between primary and secondary business objectives and inform and positively support primary and secondary business activities.

Alexander (1996) and other authors acknowledge the need to consider facilities management at four levels: corporate, strategic, tactical and operational. At a corporate level, senior managers with responsibility for facilities must contribute to service planning, formulate policy and undertake scenario planning. This requires a full understanding of the corporate culture and the levels to which responsibility and authority are devolved. Next, at the strategic level, managers carry responsibility for effective business planning of the facilities services, leadership of the team and the development of proposals for developing facilities. At the tactical level, the facilities manager ensures service quality, manages value and implements risk management strategies. The facilities manager ensures operational control through auditing and monitoring performance. Responsibilities for delivering an innovative service should be effectively delegated to service providers. Finally, at the operational level, the facilities manager is responsible for the operation and maintenance of buildings and for the delivery of the services.

Recognition of these levels of activity will explain and clarify the different interpretations of the concepts, methods and techniques of facilities management and can help to resolve confusion that might surround its development and application.

1.2.2 Workplace

In his introduction to the Symposium, Keith Alexander sought to clarify the workplace concept and to relate to discussions about the nature of workplace knowledge held at previous EuroFM conferences in Brussels (1994) and in Barcelona (1996).

For the purpose of the Salford Symposium, the workplace concept was interpreted broadly to relate considerations of 'the physical settings in which work happens, to the services that support people in those settings and, perhaps most critically, the management processes that enable their effective use'. These relationships need to be considered in the context of particular organisational contexts (culture) through cycles of time.

The concept of workplace embodies many types of work activity, not only the administrative and clerical work conventionally found in offices, but also healthcare, education and industrial production, in a variety of settings and in dispersed locations.

Becker and other authors have used the term 'total workplace' to represent these interactions, and to describe workplace strategies, planning and processes. Workplace strategies focus on harnessing the 'total workplace' to improve business performance. Workplace planning is about recognising and managing change, integrating functions within an organisation and developing intra-company relationships in an urban setting. Workplace processes also seek continuously to improve service quality by managing people, systems and information.

But what kind of facility is the workplace? To manage a facility called a workplace, it is first necessary to be clear as to what it actually is. The two components—work and place—may be considered separately.

First, work is purposeful productive activity and may involve making material goods, creating or transforming information, and/or offering a service. But there are many other types of activity that may be recognised as work: housework, child care, looking after the frail, elderly, sick; maintaining or improving the environment, such as the landscape; music, drama, dance and other forms of artistic production; and also supporting leisure and sports activity. Many authors suggest that the 21st century will be marked by breaking down the boundaries between these different productive activities, creating a much richer mix of work and other lifestyle pursuits. We are thus likely to have workplaces with child care, sports with health care, and also teaching and research with manufacture.

The second component, place, is more than space and carries the connotation of being specific. Thus one place is not the same as another. But what defines a 'place'—space, function, or form? Activity, which is the social function of space, turns it into a place. Place is a space with a specific group carrying out a specific activity (production), which in this case is a place for work. Place cannot be defined simply by physical qualities (e.g. area, environment or shape) or in terms of its contents (machines, equipment, services and furniture). It is a psychological concept and is defined in terms of cognition (knowledge) and feelings; hence, the need for greater awareness of current ideas on cognitive and affective aspects of place.

If these new types of work are undertaken in places whose qualities are defined in terms of knowledge and feelings, then managing a facility called 'workplace' requires new facilities management techniques that recognise these changes. New models of production, social relations and value-for-money need to be developed. What will they be like?

1.2.3 Action research

Keynote presentations also raised issues concerning the nature of facilities management research in order to generate discussion of appropriate methods and techniques.

Barrett referred to the construction life cycle, from concept of needs to maintenance and operation, in order to introduce related research about construction processes and to show how facilities management could benefit from a link with construction management research.

He drew on three areas of construction management research to illustrate this point. First, he identified elements of improvement and progress, and charted phases of improvement, plateau and decline in the process of continuous improvement. He then proposed a practical change model for improving performance over time, progressing through stages of tacit action, explicit understanding, mini-experiment to tacit understanding. Finally, he referred to cross-cultural technology transfer processes (Lillrank, 1995) involving abstraction, transfer and application.

The emphasis on processes—of change management, continuous improvement and technology transfer—provides a rich area for cross-disciplinary learning.

Barrett also distinguished three types of research—microscopic, telescopic and periscopic—each with a different focus and different role in relation to theory and application. Microscopic research adopts a closed systems approach to conduct abstract studies of an object's structures and causal powers, whereas telescopic research uses an open systems perspective to undertake concrete studies about mechanisms and events. Periscopic research tests the practical adequacy of solutions. Each type of research can inform the others and enable either a closer or wider focus on phenomena.

According to Barrett, research and practice in facilities management are synergistic. The essentially practical nature of the field means that much research work in facilities management is conducted through an action research approach (Greenwood and Levin, 1998, Skaret *et al.*, 2001, Alexander *et al.*, 2003), in which the researchers act together with the organisational actors in producing change related to new work forms, new offices or new systems. In this way, new knowledge is co-generated (Elden and Levin, 1991). Case studies based on observation, participation and interviews with both end users and the management in the participating organisations provide an effective vehicle of collaboration.

1.3 WORKPLACE KNOWLEDGE

Previous discussions about the nature of workplace knowledge (see Table 1.1) at EuroFM conferences, in particular the one held in Barcelona in 1996, provided a basis for further consideration at the Symposium.

Table 1.1 Workplace knowledge.

Individual and group response to environment, work situation
Human/hardware interfaces—people's understanding of, like/dislike of, attitudes to, misuse of e.g. production and communication systems, physical space and its furniture, emergency equipment (e.g. fire)
Direct and indirect costs of doing something
Direct and indirect costs of not doing something
Costs and performance of building, services, energy, information and maintenance systems
Economic and life cycle costs-cleaning, maintenance, repair
Capital costs
Land costs and investment patterns
Organisational behaviour
Decision making processes both within and outside enterprises—how are decisions made, by whom, when, how related to each other, how recorded?
Legal framework (finance, health and safety, EU and national legislation), arbitration

Source: Markus (1996)

The Symposium sought to structure and extend this earlier discussion by posing four key questions:

1. Productive workplaces—what is the impact of the quality of environment and support services on organisational effectiveness and business success?
2. Innovative workplaces—how can we assimilate new workplace technologies, processes and systems to enable flexible working?

3. Knowledge workplaces—how do we create good workplaces for knowledge work taking into account organisational issues and relations, information and communication technology and the physical layout/design?
4. Sustainable workplaces—how do we ensure that we minimise the adverse effects of the business on the environment and the local community?

An additional, overriding question was posed in the introductory session—what contribution do facilities make to organizational success and how we can identify the value added through effective facilities management?

Individual members of the scientific committee took responsibility for leading each theme and wrote introductory papers to create a context for individual contributions in order to raise key issues and to stimulate discussion.

1.4 FACILITIES MANAGEMENT FUTURES

A further symposium theme focused on issues arising out of the consideration of the future of the workplace and introduced scenarios to generate discussion about the development of facilities management as a discipline and a profession and the market to meet fie strategic needs of organisations.

At the millennium, a number of UK authors were prompted to speculate on the future of facilities management. A number of these papers were summarised at the 2002 Symposium to provide background for the discussion of future scenarios and trends.

Grimshaw and Cairns (2000) sought to identify the underlying forces influencing the global development of facilities management. They argued that radical movements in demand side organisational structures were bringing about fundamental change in the relationship between businesses and their supporting infrastructure and that, if facilities management were to generate an ability to enhance business performance via the effective application of infrastructure resources, then it would have to respond positively to this new landscape. This would involve re-evaluation of the structure that supports the development of facilities management, including the system for producing facilities management knowledge, implying new models that integrate research and practice.

In an earlier contribution, Grimshaw (1999) looked at facilities management in the wider context of the social, economic and political changes that were taking place at the end of the twentieth century. He argued that the core of facilities management relates to managing the changes that are taking place in the relationship between organisations, their employees and their facilities, all of which are being fundamentally altered by external forces. The article pleaded for a greater understanding of the issues raised by the debate around post-modernism and the impact they have had on the development of facilities management to date and will have on its future.

Varcoe (2000) drew on the results from a survey of real estate practices to identify changing trends in the property industry, and used the findings as a basis for identifying and exploring how these changes are manifested as commercial business propositions, and where this may lead the industry.

Price and Akhlaghi (1999) examined best practices in several areas of facilities management, based on case work completed over a number of years by the Facilities Management Graduate Centre of Sheffield Hallam University. The authors compared practices by reference to two dominant paradigms, or patterns, of modern organisational theory and argued that a view of organisations as living, learning systems better explains—and more importantly, better enables—best practice. The challenges that facilities managers will meet in the future are, as in other areas, those of finding new ways of leading, of cultivating environments for performing, and of finding new conversations with clients, customers and staff.

Nutt (2000) explored four generic facilities management trails to the future. These trails follow the four types of resource that are basic to the Facilities Management function; the financial resource trail (business), the human resource trail (people), the physical resource trail (property) and the knowledge resource trail (information). These four trails were considered in turn, with speculation as to the opportunities and risks that each competing future might hold. Nutt concluded with nine strategic positions from which a rich, robust and diverse variety of viable futures for facilities management can be developed.

Hinks (2000) suggested that as state-of-the-art intranet and video conferencing technologies become increasingly accessible and affordable, the possibility arises for globally dispersed virtual organisations operating in knowledge-driven sectors with little need for dedicated or permanent physical facilities. The paper questioned the role of facilities management in supporting these types of organisation and discussed the implications of e-business as a way of working for, and of, the facilities management sector.

Jones (2000) described how the UK Private Finance Initiative in the public sector has introduced radical innovations in the ways that new facilities and ongoing support services are financed. Fundamental changes in the concepts of business accommodation and service delivery are also under way, in which property is increasingly viewed as a business service rather than as a financial asset, thus helping to free core business capital while reducing the costs and increasing the quality of support service delivery. The diversification of the rules governing the property market, coupled with an increasingly sophisticated range of outsourcing arrangements, promises to provide a much greater variety and more flexible set of business support environments for the future, he concluded.

1.5 A WORKPLACE OF THE FUTURE

Haugen used examples from the recently completed Telenor headquarters at Fornebu, Oslo, to highlight facilities management challenges in managing the workplaces of the future.

Telenor's headquarters was created as a model of the workplace of the future. The vision was to create the leading innovative workplace in Scandinavia, designed to promote the sharing of knowledge and leading to the development of new workplace solutions.

The recently occupied headquarters is intended to enhance Telenor's profile to the outside world and proclaim its identity, ambitions and self-understanding to its employees. Telenor will relocate its staff from 35 to 40 different office addresses in the Oslo region to Fornebu. It was intended to give this unique area a functional, aesthetic and environmental profile to reflect Telenor's technological development based on people and nature's premises.

Business objectives were set in five main areas to guide the development: innovation, functionality, environmental responsibility, aesthetic quality and profitability. These intentions were laid out in the prospectus for the new building:

'Innovation—Telenor's headquarters will be designed to stimulate innovation and the rapid sharing of knowledge. The building will house the workplace of the future for several thousand people working in an extremely dynamic industry. This requires a great deal of flexibility on the one hand, while on the other hand also creating a sense of belonging and physical frameworks that respect each employee's individuality and integrity.

Functionality—just as modern communications technology promotes networks rather than a hierarchy, the offices will be open and democratic, receptive and varied. The building must be readily adaptable to changes in demand and framing conditions, yet enhance the company's requirements for functional flexibility.

Environmental responsibility—the construction will reflect the company's ambitions to contribute to sustainable development. The construction process will be based on keen awareness of ecological challenges, and will have a positive impact on the local environment. This philosophy will underpin the overall planning of the area and the building's location and design. Preservation and development of nature and the cultural landscape will always be kept in focus.

Aesthetic quality—as part of our efforts to promote the aesthetic in general, and architecture and design in particular. Telenor wants the company's new headquarters to reflect a high level of aesthetic quality. These aesthetic qualities will be based on what we believe Telenor stands for as a company and will reflect our positive attitude towards the location at Fornebu. The company's ambition is that the facilities will represent an architectural milestone at the start of a new millennium.

Greater profitability—reduced operating costs, increased returns on the company's intellectual capital and a profiled arena for customer contact will help strengthen the company's profitability.'

Telenor believes that success in the future will depend on creativity, imagination and innovation in the development of new information and communication technology (ICT). To Telenor, the trend in the workplace of the future is to establish arenas for communication and sharing of knowledge.

In the future, a growing number of employers will spend more time outside the traditional office. 'The office' will become more like a meeting place where you exchange ideas with colleagues, hold meetings, learn and improve the social climate of your workplace. The design of the workplace of the future must therefore enhance communication, creativity, and cooperation, and must provide arenas both for teamwork and individual concentration and, not least, the possibility for learning while working.

To ensure rapid communication and sharing of knowledge, creativity and innovation, open work areas are important. But even open-plan solutions have many sensible balances between work areas and the possibility to withdraw for private conversations and work on one's own.

Fornebu creates a variety of workplace solutions suited to various types of activity. The employees are able to use different zones in the workplace area for various purposes as the need changes. Sometimes there is a need to sit in a room completely alone. At the same time, there will be opportunities to hold different types of meetings—large or small, informal or organised.

Project work will play an increasingly larger role at Telenor. Teams will address tasks, where associates, frequently from different units, will work together for a particular period of time.

The workplace areas must therefore be so flexible that they can be readily adapted to changing needs. This will change the office as we know it today—for example, electricity, telecommunications and computer cabling will be laid in the floor with easily accessible connection points. There will be fewer permanent solid walls that require cumbersome and costly rebuilding each time a workplace area is to be changed.

Management will be increasingly a matter of having a good overview, motivating workers and being able to utilise the skills and qualities of employees. Proximity and communication between managers and staff will thus be decisive. The design of the workplace must enhance networks not hierarchy.

Various business units in Telenor were directly involved in the development of their own workplace solutions. To benefit from relevant experience, three pilot projects were established in order to test the various workplace solutions and work forms in practice.

1.6 INNOVATION AND PERFORMANCE

The keynote session at the symposium laid the foundations for developing an understanding of research in facilities management. It defined terms, introduced and clarified concepts, raised questions and offered research perspectives on the workplace, management processes and future challenges.

Papers included in the book have been selected using a three-stage refereeing process and are organised in four main sections, representing the focus and the outcomes of discussions at the symposium:

1. Organisational change and learning—theoretical perspectives on space and the physical infrastructure of work, and the development of processes that recognise and contribute to the means and management of change and learning in the context of organisations.
2. Innovation—consideration of the processes of providing function and support to sustain organisations and respond to change.
3. Performance—an overview of the major issues encountered when measuring facilities management performance and improvement in the context of organisations.
4. Towards knowledge workplaces—identification of the key workplace issues faced in meeting the challenges presented by the knowledge organisations of the future.

Altogether, there are eleven chapters with contributors from five countries: France, Malaysia, Norway, Sweden and the UK. In terms of facilities, the coverage is also broad, although many contributions have their focus on offices.

A proposal for a European Workplace Knowledge Network, originating in discussions held in conjunction with the 2002 Symposium, is found in the Appendix.

1.7 REFERENCES

Alexander, K., 1996, *Facilities management: theory and practice,* (London: E&FN Spon).

Alexander, K., Kaya, S. and Nelson, M.-M., 2003, Facilities management: a new action research approach. Paper for the BEAR 2003 CIB W89 International Conference on Building Education and Research, Salford, 9–11 April 2003.

Barrett, P., 1995, *Facilities management: towards best practice,* (Oxford: Blackwell Science).

Barrett, P. and Baldry, D., 2003, *Facilities management: towards best practice,* 2nd ed., (Oxford: Blackwell Science).

Elden, M. and Levin, M., 1991, Co-generative learning: bringing participation into Action Research. In *Participatory Action Research*, edited by Whyte, W.F., (Thousand Oaks, CA: Sage), pp. 127–142.

EuroFM, 2001, *Production Workspace: Improving the quality of production via workspace design*, Vols 0–11, (Nieuwegein: Arko).

Greenwood, D.J. and Levin, M., 1998, *Introduction to Action Research: Social Research for Social Change,* (Thousand Oaks, CA: Sage).

Grimshaw, B., 1999, Facilities management: the wider implications of managing change. *Facilities*, **17**, pp. 24–30.

Grimshaw, B. and Cairns, G., 2000, Chasing the mirage: managing facilities in a virtual world. *Facilities*, **18**, pp. 392–401.

Hinks, J., 2000, Distance is dead, long live distance? Business, virtuality, and FM in the future. In *Proceedings of the EuroFM/IFMA World Workplace Europe Conference, Glasgow, 11–13 June 2000,* (Brussels: EuroFM/IFMA), pp. 219–225.

Jones, O., 2000, Facility management: future opportunities, scope and impact. *Facilities*, **18**, pp. 133–137.

Lillrank, P., 1995, The transfer of management innovation from Japan. *Organization Studies*, **16**, pp. 971–989.

Markus, T., 1996, EuroFM Research Forum 'Developing Knowledge'. In *Proceedings of the EuroFM/IFMA Conference & Exhibition on Facility Management, Barcelona, 5–7 May 1996,* (Brussels: EuroFM/IFMA), pp. 510–512.

Nutt, B., 2000, Four competing futures for facilities management. *Facilities*, **18**, pp. 124–132.

Price, I. and Akhlaghi, F., 1999, New patterns in facilities management: industry best practice and new organisational theory. *Facilities*, **17**, pp. 159–166.

Skaret, M., Son, G. and Roberts, H., 2001, Diversity in Action Research. Paper presented at the Annual EGOS Conference, Lyon, France, July 5–7, 2001.

Varcoe, B., 2000, The possible future for Facilities Management. In *Proceedings of the EuroFM/IFMA World Workplace Europe Conference, Glasgow, 11–13 June 2000,* (Brussels: EuroFM/IFMA), pp. 43–51.

Organisational Change and Learning: An Introduction

Keith Alexander

The three chapters in Part One present theoretical perspectives on space and the physical infrastructure of the work, in the context of organisations. The papers deal with functional and symbolic aspects of space and seek to link key concepts, such as representation and experience, to social and organisational theory. In different ways, each paper argues for facilities management as a humanistic discipline, and for the development of processes that recognise and contribute to the effective management of organisational change and learning.

The papers seek to establish a theoretical basis for facilities management, concerned with the management of social space, in order to underpin. In Chapter 2, Grimshaw argues for linking facilities management to the key theoretical and political debates of the day, to support its healthy academic and professional development. He addresses important links between facilities management and social theory and discusses the relationship between the conceptual background to western spatiality and critical theory. He argues that the physical infrastructure of work is one of the most important aspects of spatial experience for many individuals and plays an active role in social processes. As a discipline, facilities management mediates between employees and employers via the physical workplace, and has a contribution to make to the development of social theory. In doing so, Grimshaw attempts to link facilities management back to the mainstream social and organisational disciplines and provide a route for facilities management to contribute to contemporary social debates.

Fenker describes in Chapter 3 a long-term, participatory action research carried out over a period of five years in a French company, exploring different phases of a process of organisational change. The research shows how divergent representations of space compromise a common understanding of goals pursued by the organisation. This analysis shows that space plays a role in organisational learning by highlighting the divergence that may exist between the different representations. It reveals how space management could reflect and support the concept of change as an iterative process of elaboration.

Fenker questions the role of facilities management within the management of organisational change by exploring, on the one hand, some recent concepts of internal dynamics, notably those based on studies of organisational learning, and, on the other hand, the relevance of a distinction between built environment and representations perceived through this environment. He proposes the management of representations as an essential contribution to the management of organisational change.

In Chapter 4, Bröchner and Dettwiler describe a project that forms part of the new national Swedish research program—'The client with the customer in focus'. Most studies of new office workspace layouts have dealt with existing large companies rather than with small and growing companies. The paper relates theories of company growth to theories of space use and relocation. Thus, theories based on stages of company development are linked here to architectural theories. The chosen indicator of growth is employment, but the relation between numbers of actual users of space and numbers of employees is expected to vary according to stage of development. Also, the symbolic importance of space varies with stages. Results are to be used for empirical surveys of how growth firms organize their facilities and facilities management.

CHAPTER 2

Space Place and People: Facilities Management and Critical Theory

Bob Grimshaw

2.1 INTRODUCTION

For its healthy academic and professional development, facilities management needs to be linked in to the key theoretical and political debates of the day. This chapter addresses the important link between facilities management and social theory. It will discuss the relationship between the conceptual background to western spatiality and critical theory. It will argue that the physical infrastructure of work is one of the most important aspects of spatial experience for many individuals and plays an active role in social processes. It will also argue that facilities management, as a discipline that mediates between employees and employers via the physical workplace, has a contribution to make to the development of social theory. In doing so, it attempts to link facilities management back to the mainstream social and organisational disciplines and provide a route for facilities management to contribute to contemporary social debates.

2.2 FACILITIES MANAGEMENT AND THE HUMANIST TRADITION

Is it too fanciful to claim that the physical workplace is one of the most important spatial regulators in modern western society and that it plays an important role in defining western mankind? If this were the case, those professions, like facilities management, that play a role in the creation and management of the working environment, would carry an important social responsibility. It would be desirable to link such professional practice to the wider debates within social theory both as a contribution to those debates and to reinforce the validity of the professional role. But in some ways facilities management originated from such a position: it can be traced back to the belief that the physical environment of work, via its impact on people in the workplace, played a major social role. Frank Becker, one of its earliest and most influential academics, came from a strong applied social science tradition and much of his early work was related to user needs in housing (Becker, 1977). His involvement with facilities management came from this root and was related to a belief that the physical environment played a crucial role in the human processes of work. This concern for the 'user' was not motivated solely by humanist concerns for the well being of people at work, but was also driven by a belief in the economic benefits of a better working environment—especially at a time of great change (Becker, 1991; Becker and Steele, 1995).

Becker's philosophy was part of a humanist tradition in organisational theory that can be traced back to nineteenth century philanthropic industrialists like Owen, Salt and Cadbury who built utopian communities for their workers (Crease, 1992). However, it remained of marginal significance until the First World War when the imperative for greater productivity to aid the national war effort led to the establishment of the National Institute for Industrial Psychology, directed by Charles Myers, tasked with examining the link between people and work (Rose, 1990). This tradition was continued by Mayo in the 1930s via the Hawthorne Experiments (Burrell and Morgan, 1979; Hatch, 1997; Cairns, 2002). It extended into the post war period with the Tavistock Institute driving forward the Human Relations movement (Locke, 1976) and the Quality of Working Life movement of the 1970s that was based on the work of Emery and Trist (Rose, 1990). The underlying theme of this tradition was higher productivity via a concern for individual well-being and needs. It sought to counter the view of employees as the dumb extension of a machine and was committed to a facilitative rather than a directive environment. The growth of environmental psychology in the 1970s carried the work forward on a broader front in an attempt to introduce the behavioural sciences to the design professions (Lang, 1987). This humanist/ behaviouralist tradition has provided a parallel if less influential strand of organisational theory to the dominant positivist, command and control, scientific management theories typified by Taylor and his predecessors (Burrell and Morgan, 1979).

In what can be regarded as a lost opportunity for facilities management, Becker failed to specifically link his position on facilities management back to the social theory of the time even though his ideas clearly came from this humanist tradition. This has deprived facilities management of a strong conceptual base in social science that would have benefited those who have followed Becker's lead in developing the social and behavioural strands of facilities management (Cairns, 2002; Grimshaw, 1999; Grimshaw and Cairns, 2000; Haynes and Price, 2002; McLennan and Cassels, 1998)—in many ways this chapter is an attempt to set this record straight and re-make the links between the social aspects of facilities management practice and social theory that are still missing.

This is necessary because of the persistence of the functionalist organisational culture that consistently adheres to the view that the physical environment of work plays little if any role in the social interactions we define as work. Cairns (2002) blames this persistence on a wilful misinterpretation of the outcomes of the Hawthorne experiments that 'separated the physical and social environments of work, with the social as the primary area of study'. Cairns goes on to say that this split was 'reinforced by Herzberg's placing of the physical within the dichotomous framework of motivating/hygiene factors'. To Herzberg *et al.* (1959) the physical environment was not a key motivating factor for most workers. Even the more balanced view formulated by Locke (1976) concludes that 'physical conditions [...] unless they are extremely good or extremely bad, are usually taken for granted'. Today it is still noticeable in the literature on general organisational theory that most research on the 'working environment' ignores the physical infrastructure. Amabile's work on creativity is a case in point; she is the leading researcher on the link between working environment and creativity but her definition of the working environment does not include the physical environment which she takes to be neutral (Amabile, 1993). However, Hawthorne and Herzberg are just more recent manifestations of a more deep seated problem in western philosophy's attitude to space and utility.

Lane (2000) refers to the entrenched economic dogma, especially in the USA, that denies 'the belief that many of [people's] main pleasures in life come from their work' and that work, being defined as a disutility, is merely the 'pain necessary to earn the pleasures of money and leisure'. This essentially Benthamite doctrine denies that work can be an end in itself and holds that 'aversion not desire is the only emotion which labour taken by itself is qualified to produce' (Stark, 1952). In being thought of as what Cludts (1999) terms the 'rational utility maximising individual', the modern employee is motivated only by the need to earn money. In this prevailing culture the physical infrastructure is provided solely to support task efficiency and there can be no link between user needs and productivity.

By contrast, Forty (2000) points to Kant's specific rejection of utility as an element of aesthetics. Forty argues that this has detached architectural theory and practice from any concerns with user needs. He reports that with a few notable exceptions this rejection has prevented architects and others involved in spatial design from responding to user needs as an integral part of their design ethos. The inability of the leading design profession to recognise utility as a key driver of its work has undoubtedly influenced wider organisational thinking on workplace provision. It has certainly skewed the development of all the professions related to the built environment away from user concerns.

Taken in total this strong rejection of the impact of the working environment on social processes has affected the development of facilities management. It lies behind Duffy's (2000) contention that facilities management has been pushed to direct itself towards outsourcing and cost driven measures, forcing facilities management practice into a narrow 'cost cutting function'. Its early concerns with 'user needs' and the benefits of a better 'fit' that linked facilities management via productivity to strategic planning is in danger of being lost. This failure has effectively excluded facilities management from any contribution to the organisational debates about what adds value to the work process. Facilities management has failed to demonstrate the value-added element of the physical environment and consequently makes little contribution to organisational strategy.

2.3 WORKSPACE AS PLACE

Yet on wider reflection this functionalist position is not tenable. Western industrial society is a spatially focused society that has constantly used physical structures to both mirror and reinforce its culture at all levels of society (Dovey, 1999). French philosophers Deleuze and Guattari (1987) have characterised modern western society as striated, where all space is divided up, has clear patterns of ownership and has clear boundaries. Space is a private commodity that can be traded like any other commodity, and it is hard to see how our society and the economic system it is based on would work if this were not so. Utility is implicit in the view of space as a commodity. The opposite is the 'smooth' space characteristic of nomadic peoples where there are no specialist spatial functions, no individual ownership and no boundaries. This highlights the fact that many of the mechanisms that govern the social and economic relationships in western society are spatially based. Because of this space takes on a central social meaning. As Hillier states, buildings are 'not simply a background to social behaviour—[space] is itself a social behaviour. Prior to being experienced by subjects (users) it is already imbued with patterns which reflect its origins in the behaviours through which it was created' (Hillier, 1996). Far from being passive and inanimate, space is an integral part of dynamic social intercourse where the very configuration of space is part of the social information process.

Hillier, as one of the few architects advocating that spatial design has a social derivation, comes from a tradition that can be traced back via Lefebvre (1991) to the dialectical approach of Hegel and Marx. Lefebvre analyses the complex relationship between social groups of all types and the space they appropriate. In *The Production of Social Space*, his general thesis is that the process of producing social space is an integral and vital process in the development of any society or social group. The physical space a group appropriates as part of the process of formation represents the group and the group cannot be understood without reference to it. Without the link to the social processes that produce and use a space, the space itself has no meaning. The social code inherent in the space is complex and 'involves levels, layers and sedimentation of perception, representation and spatial practice which pre-suppose one another'. The occupied space gives direct expression to the relationships on which social organisation is founded and all spaces are the result of historical modes of production that integrate the social with the economic and the political. Lefebvre's purpose, according to Shields (1999), was to develop a concept of space 'from which to launch a critique of the denial of individual and community rights to space' to reflect the appropriation conflicts that are inherent in spatial relationships.

In this model, organisational space is no different from any other space and must represent in many different ways the culture and form of the social processes that make up the organisation. The integration of space and cultural meaning links work space to the rich strand of literature on place. 'Place' is physical space plus the complex and integrated social, economic and political meanings that the space carries for one or more social groups. Special places have always carried great significance in human society and are often linked to social ritual. Casey (1997), in tracing the historical thread of place in western philosophy, emphasises the mechanisms of appropriation that lead to the 'implacement' of some groups and the 'displacement' of others that are fundamental to the operation of our striated world. Casey sees work space as a 'place' where complex mechanisms are expressed in the physical environment to 'implace' individuals and groups within the organisational culture. Equally signals are given about the 'displacement' of those who are seen to be outsiders. Space reflects the culture and 'place' underlines Lefebvre's view that social change cannot happen without spatial change.

Whilst this spatial ordering of society is central to all social transactions, it would not be fair to claim that space itself forms a substantial element of social theory. A few theorists like Foucault (1979; 1980; 1988) and Castells (1996) have followed Lefebvre in putting physical space at the centre of their work; however, references to space crop up sufficiently frequently in the work of other authors to suggest that its role in social development is accepted. The contributions made by

Hetherington (1998), Jameson (1992), Soja (1996), Sennett (1997; 1998), Bauman (2000) and Beck (2000) suggest that those involved in the design and management of space play a more important role than cost based operational support.

2.4 CRITICAL THEORY

Whilst the linking of facilities management to social theory is a necessary step in its development, social theory is not a coherent body of ideas but a collection of very diverse approaches. Which is the most appropriate for facilities management? Burrell and Morgan (1979) have mapped these diverse approaches using a matrix determined on one axis by the spectrum from objectivity to subjectivity and on the other by the spectrum from the sociology of radical change to the sociology of regulation (see Figure 2.1). The matrix divides into four quadrants or paradigms that are labelled 'radical humanism' (individualistic and subjective), 'radical structuralism' (individualistic and objective), 'functionalism' (regulatory and objective) and 'interpretism' (regulatory and subjective). Into these four quadrants Burrell and Morgan then map the different strands of social theory.

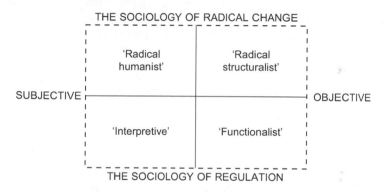

Figure 2.1 Four paradigms for the analysis of social theory.
Reproduced by permission of Ashgate Publishing from *Sociological Paradigms and Organisational Analysis,* 1979, by G. Burrell and G. Morgan.

It is clear that traditional organisational theory is firmly linked to the objectivist area of the functionalist paradigm, characterised as it is by high levels of regulation and objectivity supporting the hierarchical top down approach that seeks a single 'right' way of doing things. This is an essentially rationalist position and encompasses much of the history of the main strand of organisational development from Adam Smith's division of labour to Taylorism, Fordism and beyond. It could also be said to be the right location for current operational facilities management practice, which is seen by many as a function of top down organisational policy and is an instrument of senior management policy (Donald, 1994). This is hardly surprising, as facilities management has been a demand led function that must serve the needs of its masters to become established.

However, the ideology that underpins Becker's ideas about facilities management never did fit into this functionalist paradigm; his is essentially humanist in nature, deals with radical change, and expresses concern for the users of workplace environment. It clearly fits into the 'radical humanist' quadrant being individualistic and subjective. It is individualistic in advocating looking after the needs of individual users and subjective in that it acknowledges that different solutions suit different people and different organisations. Taking user needs into account also leads to a more dialectic approach to decision making that attempts to integrate the traditional top down with a more participatory bottom up approach (Grimshaw and Garnett, 2000). It cannot be said, however, that these ideas are 'over-radical' in sociological terms and they fall within the inner area of the paradigm into which Burrell and Morgan place critical theory.

In a stable situation, the dislocation between the intellectual origins of facilities management and the reality of its current practice would not be tenable and would strangle its onward development. There are already signs that facilities management is in danger of becoming marginalised in the way that organisations define its function. Duffy's (2000) comments on facilities management practice being driven purely by 'cost cutting' is symptomatic and illustrates facilities management's decline from its original ideals into a peripheral operational function. This relentless logic of functionalism has the potential to downgrade facilities management. But the current situation is not stable—it is exactly the opposite in many ways. In terms of organisational development, a whole raft of management theorists are speaking with one voice and saying that, in a fast changing global society driven by ICT, the functionalist paradigm as a basis for organisational governance is no longer tenable. In Senge's (1997) words 'it is no longer possible to figure it out from the top'. Senge (1997), Drucker (1997), Handy (1997), Argyris (1998), Bartlett and Ghosal (1995), Sawhney and Parikh (2001), the list goes on and on, are all exploring the ramifications of constant change and the need for the whole organisation to be creative and innovative. (Incidentally not one of these authors mentions the physical environment of work!) Their ideas, which are based on the imperative to reduce regulation and empower the individual to be creative and innovative, are also clearly within the scope of critical theory. Becker is part of the same intellectual tradition and he conceived of facilities management as a function of the same general shift that needs to take place to move organisations into a much more inclusive, participatory environment and away from its functionalist straightjacket. This shift will not be easy and Burrell and Morgan have shown that the functionalist and radical humanist paradigms are more or less diametrically opposed.

A few have followed Becker's humanist lead in trying to develop the academic underpinning of facilities management. There is the work being carried out by Cairns and Beech (1999) at Strathclyde University, intending to establish user-based workplace design and management as a serious strand of management theory; the work of Dovey (1999) in Melbourne, linking physical infrastructure to symbols of corporate power; and the work of Markus (1993; see also Markus and Cameron, 2002), also from Strathclyde, that links the physical environment to social power structures. Contributions have also been made by, amongst others, Haynes and Price (2002) and Grimshaw (1999; 2001a; 2001b). This body of work represents the only direct links between facilities management and social theory but provides the potential for future engagement with those sociologists and psychologists who have involved themselves with change and the workplace (Bauman, 2000; Beck, 2000; Sennett, 1998; Reich, 1991).

All these links fit comfortably within the scope of critical theory. Critical theory is a social philosophy whose 'proponents seek to reveal society for what it is, to unmask its essence and mode of operation and to lay the foundations for human emancipation through deep seated social change' (Burrell and Morgan, 1979). It originated in the 1930s and derived from the work of Marx and Freud. It sought to describe and explain the fundamental and negative changes in western society in that decade and was especially critical of rationalist dogma that, far from promoting the rights of individuals, was leading to more and more extreme forms of repression. As such it has always been anti-positivist and embraces difference and diversity. Geuss (1981) has described critical theory as a theory that provides a guide for human action, is inherently emancipatory, has a cognitive content and, unlike scientific theory, is self-conscious, self-critical and non-objectifying. It seeks to explain why people submit to systems of collective representation that reflect existing power structures that do not serve their objective interests. It is essentially humanist and individualistic being driven by a belief in democracy and the rights of the individual to freedom and to realise their full potential. It seeks to synthesise philosophy and social science. It is interdisciplinary in nature and connects economics, the state, society, culture and individual experience, using mainly qualitative theories and methods. It does not embrace a single theoretical position but seeks more to explain contemporary phenomena in whatever guise. It is not intended to be passive but seeks to promote social change.

In the post-war period, critical theory has been dominated by the Frankfurt School led by Horkheimer, Habermas, Marcuse and Adorno. It was eclipsed to some extent in the 1970s and 1980s by post-modernism, but because of the latter's rejection of macro themes, critical theory, which is aimed at understanding complex social changes on a macro basis, has undergone a revival (Best and Kellner, 1991).

With its central theme of the dynamic tension between the rights of the individual citizen to freedom and self-determinism and the tendency for the rationalist nation state and its institutions to impose control over its citizens, critical theory is peculiarly suited to today's situation. This is more so when the impact of globalisation and the power of global organisations is added to the equation. The need to understand these forces and their impact on the individual are greater than ever. The adaptability of critical theory has led it away from the overtly political and socialist agenda of the 1960s against the power of the corporate state in the post-war years to a greater concern for the impact of globalisation on both the individual and the nation state (Best and Kellner, 1991). This reflects the fundamental shift in global power structures that underlies the rise of corporate dominance. Indeed Castells (1997) has argued that the nation state relies for its integrity on the control of both space and time, and both he and Bauman (2000) believe that nation states are losing their grip on space and time as they lose control of capital flows and information.

The globalisation of business and the shift of power from national governments have created a new and more complex situation for the individual (Castells, 1997). Not only is there a new tension between the individual (both as employee and consumer) and the power of corporate business but there is also the effect on the individual of the tension between the nation state and global business—which may be pulling in different directions with the individual caught in the middle. The battleground for this new concern for individual rights is the workplace. And in the workplace the individuals do not have the same democratic rights to participate as they do in political decision making and they are in a much weaker position (Grimshaw and Garnett, 2000). Reports of new and more sophisticated means of control are legion (Rosenberg, 1999). In this situation all those involved in the provision and management of the workplace have a responsibility to understand the complex interactions and the impact they have on individuals. Facilities management is no exception.

Within the scope of critical theory, the issue for those involved in the provision of facilities is whether or not the physical environment of work can play a positive role in the social interactions that we define as work, or whether it is just a neutral backdrop. If the latter is the case then the physical workplace is a cost element only and one to be minimised. However, if the former is the case then the workplace becomes not just a stage for power struggles but rather an integral part of the social process. Facilities management needs to understand much more about these relationships and the implications of different approaches to workplace provision. The description of critical theory by Kellner (1995) as 'the art of making connections and discerning contradictions' seems peculiarly appropriate for facilities management which has constantly to balance the demands of the organisation with the needs of the individual within the constraints of the physical environment.

2.5 ALIENATION: ENGAGEMENT WITH CRITICAL THEORY

Once it is acknowledged that critical theory is the relevant home for facilities management in social theory then the concepts used in its application can be brought into play. One of these central concepts is that of alienation—or 'that which divorces mankind from his true self and hinders the fulfilment of his potentialities as a human being' (Burrell and Morgan, 1979). Gramsci concentrates on order, authority and discipline as propagated through nationally based institutions like the family, the school and the workplace (Forgacs, 1988). The Frankfurt School's leading thinkers have all developed different but parallel perspectives on aspects of alienation. Marcuse (1964) develops ideas around the alienating role of technology, science and logic, and Habermas (1984) has developed a theory of communicative competence built on the alienating role of language—what he terms communicative distortion. The key concept that unifies this work is an understanding of the power relationships between the individual and society.

But alienation is essentially a Marxian concept linked to the alienating features of early capitalism. His concept had elements of both the physical and abstract. The former in that workers were separated from the products they were making in a process of the division of labour that kills creativity and identity with the product; the latter in the separation of human beings from their human essence via over-controlling management practices (Shields, 1999). Both referred to the negative effects of some of the new social and economic structures that were emerging in the first phase of industrialisation. Alienation was a measure of the estrangement of people from their human rights and reflected the view that the treating of people as dumb units of production was counter to the beliefs of the Enlightenment. It would not be unreasonable to argue that part of the system Marx was commenting on was enforced by the physical facilities of the factory system and the physical conditions in new industrial cities. This being so, the physical environment of work has always been part of the problem of alienation and has the potential to reinforce the same message now.

Marxian alienation was closely linked to the dynamics of the workplace and is a wholly appropriate concept for facilities management. This view has been reinforced by Lefebvre (1991) who emphasised that the underlying aspects of alienation, related to objectification and externalisation, have a strong spatial element and that spatial arrangements have the power to include or exclude (Shields, 1999). Ross (1996), in commenting on Lefebvre's work, states that the 'dilemmas of alienation highlight the twin poles of location and identity; to be alienated; to be displaced from oneself; to be foreign to oneself'. Rose (1990) has emphasised the view that the relationships inherent in capitalism are still conflicting and almost by definition have to alienate the workforce if the system is to work. He feels that work is still made up by 'the elements of obedience, self-denial and deferred gratification' and entails a 'subordination of the subjective'. Workers still work at the 'behest of others in a process they do not control to produce goods or services that they do not enjoy'. Part of this framework of control is the physical environment.

Alienation provides a useful overarching framework for two of the key issues currently facing facilities management in the workplace:

- The move to achieve greater productivity in routine service jobs by reinforcing a culture of control by using technology.
- The expulsion of workers from the workplace into a more flexible environment where supported workspace is not provided and whereby by people are co-located in time but not in space.

These reflect a growing bifurcation in the treatment of employees in global business. Both Castells (1996) and Reich (1991) have noted the emergence of two different types of workers. Reich has labelled these the 'symbolic analytic' (or knowledge) workers who are high value adding and operate within a global market for their skills, and the 'routine production' workers whose jobs are routine and can be exported without difficulty. The workplaces needed by each group are driven by different criteria. The former (henceforth known as 'knowledge workers') need an empowered workplace where their individual skills in creativity and innovation can be exploited and which encourages them to stay. The latter (henceforth called 'routine workers') are driven by cost considerations because the work can and will be exported to the most beneficial cost locations. The flight of global manufacturing to the Far East, especially China, is indicative. Both knowledge workers and routine workers face situations with the potential to isolate individuals from social contact in some way and to lead to alienation. These situations raise serious issues about the relationship between the individual and the host organisation which includes the supporting physical infrastructure.

Although this general division of global workers into two categories is crude it does provide a useful classification for considering the strategic challenges facing facilities management. The issues raised are complex; in this short chapter it is not possible to deal with the issues raised in depth but an outline can be given by examining the following inter-linked pairings:

- Power and powerlessness
- Inclusion and exclusion
- Communication and distortion

The intention here is not to discuss the ideas of critical theory in depth but to show briefly how each of these pairings are linked to the two key problems facing facilities management.

2.6 POWER AND POWERLESSNESS

At the heart of the concept of alienation is the embedded power structure within any social situation. Organisations are social structures and their operations have an underlying power structure. There is strong support from the literature that the physical environment not only reflects the power structures but *has* to reflect them if the power structures are to function. It is significant that Foucault (1979) chose Bentham's Panopticon, the supreme example of how the physical environment can be configured to enforce control, as his metaphor for the modern organisation. This underlying belief in institutional systems being reinforced by the physical nature of the institutions underlies Foucault's work on prisons and is extended by Markus (1993) to include all emergent building types in the nineteenth century including workplaces. Markus demonstrated in each case that the power structures of the social regimes were directly reflected in the physical environment. He uses Hillier's space syntax model, which is based on the premise that layout of buildings controls social contacts and defines the nature of those contacts, to reinforce his message (Hillier and Hanson, 1984).

Dovey (1999) also uses space syntax to demonstrate a direct link between corporate power structures and the physical environment in his study of corporate office towers in Melbourne. On a wider basis, Laing, Duffy *et al.* (1998) link changes in the working environment to the need to support knowledge workers and have developed a typology of workspaces to match. Becker and Steele (1995) also argue that creative and innovative workers cannot function without changes to the physical environment to reflect new power relationships. Baldry *et al.* (1988) in using the slogan 'Bright satanic offices', and Rosenberg (1999) have shown how the physical environment of offices combined with information technology have been devised to reinforce control.

For routine workers, powerlessness has always been conveyed to some extent by the regimentation of the working environment in both manufacturing and the office (Sundstrom, 1986). But the pressures of cost control brought about by global competition is leading to more comprehensive means of control of the work process, symbolised by increased electronic monitoring and other neo-panoptical control strategies (Baldry *et al.,* 1998; Rosenberg, 1999; Rose, 1999). The eminent exportability of this type of work will always leave routine workers relatively powerless as they are unable to counter the need to keep the cost down. The cost of workplace provision is a key ingredient in the overall cost equation and will maintain cost as the key driver of workplace design.

Knowledge workers, on the other hand, may have escaped this level of direct control by dint of the high value added of their skills. But in some ways those who work flexibly and lead a nomadic existence have lost power because of a lack of fixed reference points in their lives. It has already been demonstrated that concrete symbols are important in conveying power and the prolonged detachment of many from familiar workplace environments and the social symbolism they convey may leave them detached from power networks (Sennett, 1997). There is a desperate need for a better understanding of the necessary 'implacement' mechanisms for the flexible and virtual workplaces.

2.7 IDENTITY AND EXCLUSION

Work is so embedded in our 'modern' culture that it is sometimes difficult to achieve an objective view. Social theorists like Bauman (2000) and Beck (2000) have claimed that work status has become the main prop in individual identity. Ulrich Beck (2000) sees work as being so entrenched in our culture that it is 'part of the modern European's moral being and self-image'. Work defines the person and being without work is seen as socially unacceptable.

Castells (1996) describes individual identity as a 'source of meaning and experience' and argues that identity is a result of a 'process of self-construction and individuation'. Identity is framed by the social context of the time but only becomes meaningful when this context is internalised and given meaning by the individual. Castells also argues that for most people identity is made up of many aspects of life (roles) but is usually 'organised around a primary identity', that frames the other roles and is self-sustaining across time and space. Many would argue that in the modern world it is work that has been the primary focus that frames everything else. Beck (2000) has described work as the 'only relevant source and valid measure of human beings and their activity'. This reference to the growing centrality of work in western society only serves to emphasise that the workplace is playing an increasingly central role in the lives of the working population.

Work is a complex social activity with many aspects to its make up but it would be hard to argue that the physical environment of work does not play some role in the equation. If the workplace is the primary environment for social identity those who deal with its design and management play an important role in ensuring that the physical environment of work has a positive rather than negative effect on the identities of those who use it. If the workplace is the physical manifestation of social identity then changes to the workplace will have serious social implications. The issue of identity and isolation emphasises the human need to be included within the social group. But the forces behind globalisation and manifested in the workplace are blurring the physical boundaries of work and in many respects making exclusion more common. Castells' (1996) network organisations do not have the same boundaries as the traditional organisation and who is included and how they are included is far from clear.

In terms of routine workers, although they are part of an inclusive environment, there are threats to their individual identity in the regimentation and control of the workspace that effectively excludes them (Reich, 1991). Baldry *et al.* (1998) have commented on the implications for the routine office worker with their analysis of contemporary neo-Taylorist workplaces where controls are enforced by a combination of management techniques, IT monitoring and spatial configuration that make the working environment central to the alienation process. The attendant de-skilling of work and lack of control by the individual leads to environments of total alienation that lack any symbols of individuality. This may not be the most enlightened way to maintain productivity but it is driven by the imperative to control costs if these jobs are not to be exported. Call centres have come in for particular criticism (Cairns and Beech, 1999; Markus and Cameron, 2001) but the reality is that many so-called high tech jobs are low skilled and can be re-located with ease. The transitory nature of these jobs threatens to exclude routine workers from society by lack of continuity in their working lives and ultimately via unemployment. This demonstrates the potential for alienation by lack of ability to influence global forces that leads to a lack of security and feeling of impermanence in the workplace.

Knowledge workers are not much better off. The threat to their identity is more related to the social isolation of constant moving that many corporate workers face and their inability to put down local roots. The numbers of people in flexible working environments is growing rapidly, especially in the USA. Caldwell (2002) reports that 28 million people in the US telecommute in some form and of those over 20 per cent work at home. They are potentially excluded by a lack of visibility and the symbols of permanence conveyed by space in the office. Sennett (1998) has highlighted that knowledge workers operate in a global market and are increasingly mobile in the course of their working lives. This may be economically beneficial but the impact on the social development of them and their families can be dire. They lack the basic attachments to 'place' that settled populations enjoy and suffer from social isolation and exclusion. They clearly fall within Casey's (1997) definition of displacement—their workplace is a temporary node in the flow of work and people around the all encompassing network.

The potential for exclusion for both knowledge and routine workers is a growing social problem and undermines the traditional role of the workplace as a haven of security (Beck, 2000).

2.8 COMMUNICATION AND DISTORTION

Communication using a recognised language is the fundamental driving force of any human society. At times of radical change the language can change and the establishment of a new language can be a painful and divisive process. In terms of organisational change, Handy (1997) notes the need for a new vocabulary to match the fundamental changes going on and that 'the old language of property and ownership no longer serves us in modern society because it no longer describes what a company is'. Thrift (1997) also detects the emergence of a new language of business with the emergence of new concepts like networks, soft capitalism, light structures and teamwork.

Hatch (1997), one of the few organisational theorists to consider the importance of the physical infrastructure to organisational development, has argued that the most important aspect of the relationship between an organisation and its physical infrastructure lies in the latter's symbolic qualities. On a different tack Markus (1993) in his study of the development of institutional environments in the nineteenth century uses the idea of buildings as language, packed with meaning and symbolism, as his framework for analysis. More recently both Markus and Cameron (2001) and Forty (2000) have used linguistic approaches to analyse the meaning of buildings. Markus and Cameron used discourse analysis to evaluate the underlying attitudes to buildings by studying the language used about them. Similarly Forty, in looking at the way architecture has been described over time, draws out the underlying attitudes across a range of settings and groups. All three authors emphasise that buildings are a means of communication and that the language we use about them links them to our culture.

If we think of the physical infrastructure of work as an important part of the cultural language of an organisation, then a firm link can be established with critical theory. Language and semiotics has always been has always been an important part of critical theory and perhaps the most important exponent of language and communication has been Habermas (1984). From his work Habermas concluded that 'the problem of language has replaced the traditional problem of consciousness' and he developed a theory of communicative competence to conceptualise the problem. Within this theory, language is the driving force behind 'symbolic interaction' and he contrasts the 'ideal speech situation' with that of 'communicative distortion'. Work, one of his main exemplars, is seen as an example of communicative distortion because the language system reflects the unequal power relationships within the working environment. Several authors (Markus and Cameron, 2001; Dovey, 1999; Hatch, 1997; Cairns and Beech, 1999; Grimshaw, 2001b) would support the view that the physical infrastructure is part of the organisational language and does reinforce the power structures within the organisation. It is therefore part of the communicative distortion process that restricts individual rights for the benefit of the organisation as a whole.

For routine workers the highly controlled and depersonalised forms of communication that are imposed by the company reinforce their alienation. This communication is reflected in the working environment in conditions that Baldry *et al.* (1998) describe as Dickensian. 'If we combine office workers' experience of work intensification under Team Taylorism with their daily ordeal at the mercy of a malfunctioning built environment we can see that the total reality does not seem "modern" at all but almost approximates to a (19th century) sweatshop'.

Knowledge workers have the opposite problem. Their lack of face-to-face communication and the substitution of less personal forms have distorted the quality of communication. Their lack of symbolic markers or a secure base from which to protect their communication means they are always exposed. Even where they are provided with a custom made 'knowledge' environment the lack of communication between management and employees can lead to a catastrophic alienation as has been demonstrated by the Chiat Day debacle (Berger, 1999). Their innovative and high profile office design in Los Angeles proved to be completely unworkable. In the words of the author 'It was a bold experiment in creating the office of the future. There were no offices, no desks, no personal equipment. And no survivors'.

2.9 A CRITICAL THEORY OF FACILITIES MANAGEMENT

Alienation is a useful concept for facilities management in that it provides a framework which stimulates a better understanding of the complex social interactions facilities management is dealing with. It is clear that the above pairings of concepts reflect the alienating potential of rapidly changing organisational circumstances and that facilities are a significant factor in each dialectic. Facilities are a central part of the power relationships that govern work; they play a key role in the symbolism that underpins individual identity conveyed by work; they are important in conveying a sense of either social inclusion or exclusion; and they are a part of the communication structure of the working culture. facilities management is in the front line of countering this alienating potential for both key groups of workers. It is part of a movement led by organisational theorists against the domination of the functionalist paradigm. If organisations have to be flexible, creative and innovative to survive, this can only be achieved if the design and management of the working environment is embedded in the change process.

From this and the above, a number of propositions can be stated that provide a link between facilities management and critical theory:

- The management of change in the working environment is primarily a social task.
- The physical infrastructure of work adds value by its impact on social interactions. It is part of the organisational culture and has the potential to alienate or attach.
- Facilities management as a service should be focused on issues of attachment and alienation that underlie motivation and productivity.
- Facilities management can make a positive contribution to human fulfilment and self-esteem by promoting a physical environment that stimulates creativity and innovation.
- Facilities management subscribes to the central tenet of critical theory in that it is humanist, non-objective and aims to shape the future and not just report on it.

The link with critical theory outlined in these propositions is not prescriptive or driven by a single dogma but is qualitative and flexible. It grounds in social theory the claim of facilities management to be a humanist profession with a credible social responsibility for people in the workplace (Grimshaw, 2001b). In the polarised workplace, where some groups are under tighter traditional control and others are displaced from the workplace altogether, the concepts of power, identity, inclusion and communication provide the dialectics to gain a better understanding of the consequences of different facilities management strategies in preventing detachment of individuals and groups from both the organisation and society as a whole.

This chapter has shown that facilities management can engage with critical theory. Facilities management is part of a process that links individuals to an organisational culture via the configuration and management of the physical infrastructure and its symbolic dimensions. It seeks to understand why and how people buy in to such a culture and what compromises they have to make in physical space to do so. Facilities management can be passive and merely an instrument of central management policy—but at its best, and as conceived by Becker, it can be an active agent of change that integrates its technical, economic and social skills to promote social change for the benefit of all. It can contribute to the onwards development of social policy in one of its most important and problematic arenas—the workplace.

2.10 REFERENCES

Amabile, T.M., 1993, Motivational Synergy: toward new conceptualisations of intrinsic and extrinsic motivation in the workplace. *Human Resource Management Review*, **3**, pp. 185–201.

Argyris, C., 1998, Empowerment: the emperor's new clothes. *Harvard Business Review*, May–June.

Baldry, C., Bain, P. and Taylor, P., 1998, 'Bright Satanic Offices': intensification, control and team Taylorism. In *Workplaces of the Future*, edited by Thompson, P. and Warhurst, C., (Basingstoke: Macmillan Business).

Bartlett, C.A. and Ghoshal, S., 1995, Changing the Role of Top Management: beyond systems to people. *Harvard Business Review*, May–June.

Bauman, Z., 2000, *Liquid Modernity*, (Cambridge: Polity Press).

Beck, U., 2000, *The Brave New World of Work*, (Cambridge: Polity Press).

Becker, F.D., 1977, *Housing Messages*, (Stroudsberg, PA: Dowden, Hutchinson and Ross).

Becker, F.D., 1990, *The Total Workplace: facilities management in the elastic organisation,* (New York: Van Nostrand Reinhold).

Becker, F.D. and Steele, F., 1995, *Workplace by Design: Mapping the high performance workplace,* (San Francisco: Jossey-Bass).

Berger, W., 1999, Lost in Space. *Wired*, **7**(2), pp. 76–81.

Best, S. and Kellner, D., 1991, *Postmodern Theory: critical interrogations*, (London: Macmillan).

Burrell, G. and Morgan, G., 1979, *Sociological Paradigms and Organisational Analysis*, (Aldershot, Hants: Ashgate).

Cairns, G., 2002, Aesthetics, Morality and Power: design as espoused freedom and implicit control. *Human Relations,* **55**(7), pp. 799–820.

Cairns, G. and Beech, N., 1999, Flexible Working: organisational liberation or individual straightjacket. *Facilities*, **17**(1/2), pp. 18–23.

Caldwell, F., 2002, Creating Resiliency with the E–workplace. Gartner Research.

Casey, E., 1997, *The Fate of Place: a philosophical history,* (Berkeley, CA: California University Press).

Castells, M., 1996, *The Rise of the Network Society,* 2nd ed., (Oxford: Blackwell).

Castells, M., 1997, *The Power of Identity*, (Oxford: Blackwell).

Cludts, S., 1999, Organisation Theory and the Ethics of Participation. *Journal of Business Ethics*, **21**(2), pp. 157–171.

Crease, W.L., 1992, *The Search for Environment*, (Baltimore: John Hopkins University Press).

Deleuze, G. and Guattari, F., 1987, *A Thousand Plateaus*, (Minneapolis: University of Minnesota Press).

Donald, I., 1994, Management and Change in Office Environments. *Journal of Environmental Psychology*, **14**, pp. 21–30.

Dovey, K., 1999, *Framing Places: Mediating Power in the Built Form*, (London: Routledge).

Drucker, P.F., 1997, The Future That Has Already Happened. *Harvard Business Review*, Sept–Oct, pp. 20–24.

Duffy, F., 2000, Design and Facilities Management in a Time of Change. *Facilities*, **18**(10/11/12), pp. 371–375.

Forgacs, D., 1988, *A Gramsci Reader*, (London: Lawrence and Wishart).

Forty, A., 2000, *Words and Buildings*, (New York: Thames and Hudson).

Foucault, M., 1979, *Discipline and Punish,* (New York: Vintage Books).

Foucault, M., 1980, *Power/Knowledge,* (New York: Pantheon Books).

Foucault, M., 1988, *The Care of the Self,* (New York: Vintage Books).

Geuss, R., 1981, *The Idea of a Critical Theory: Habermas and the Frankfurt School*, (Cambridge: Cambridge University Press).

Grimshaw, R.W., 1999, The Wider Implications of Managing Change. *Facilities*, **17**(1/2), pp. 24–30.

Grimshaw, R.W., 2001a, Ethical Issues and Agendas. *Facilities*, **19**(1/2), pp. 43–51.

Grimshaw, R.W., 2001b, Facilitating Work: FM and people in new organisations. In *Proceedings of Ideaction 2001 People Creating Value, 12th National Conference of the Facility Management Association of Australia Conference, Melbourne.*

Grimshaw, R.W. and Cairns, G., 2000, Chasing the Mirage: managing facilities in the virtual world. *Facilities*, **18**(10/11/12), pp. 392–401.

Grimshaw, R.W. and Garnett, D., 2000, Workplace Democracy. In *Facility Management: Risks and Opportunities,* edited by McLennan, P. and Nutt, B., (Oxford: Blackwell Science), pp. 107–116.

Habermas, J., 1984, *The Theory of Communicative Action,* (Boston: Beacon Press).

Handy, C., 1997, The Citizen Corporation, *Harvard Business Review*, Sept–Oct, pp. 26–28.

Hatch, M.-J., 1997, *Organization Theory: Modern, symbolic and postmodern perspectives*, (Oxford: Oxford University Press).

Haynes, B. and Price, I., 2002, Quantifying the Complex Adaptive Workplace. In) *Proceedings of the CIB Working Commission 070 Facilities Management and Maintenance 2002 Global Symposium, Glasgow*, edited by Hinks, J., Then, D.S.-S. and Buchanan, S., (Glasgow: CABER, Glasgow Caledonian University), pp. 403–416.

Herzberg, F., Mausner, B. and Snyderman, B.B., 1959, *The Motivation to Work*, (New York: John Wiley & Sons).

Hetherington, K., 1998, *Expressions of Identity: Space, Performance, Politics*, (London: Sage Publications).

Hillier, B., 1996, *Space is the Machine: a configurational theory of architecture*, (Cambridge: Cambridge University Press).

Hillier, B. and Hanson, J., 1984, *The Social Logic of Space*, (Cambridge: Cambridge University Press).

Jameson, F., 1992, *Postmodernism or the cultural logic of late capitalism*, (Durham, NC: Duke University Press).

Kellner, D., 1995, *Media Culture: cultural studies, identity and politics between the modern and the postmodern*, (London: Routledge).

Laing, A., Duffy, F., Jaunzens, D. and Willis, S., 1998, *New Environments for Working*, (London: BRE Publications).

Lane, R.E., 2000, *The Loss of Happiness in Market Democracies*, (New Haven, MA: Yale University Press).

Lang, J., 1987, *Creating Architectural Theory: the role of the behavioural sciences in environmental design*, (New York: Van Nostrand Reinhold).

Lefebvre, H., 1991, *The Production of Space*, (Oxford: Blackwell).

Locke, E., 1976, The nature and causes of job satisfaction. In *Handbook of industrial and organizational psychology*, edited by Dunnette, M.D., (Chicago: Rand McNally), pp. 1297–1349.

Marcuse, H., 1964, *One Dimensional Man*, (Boston: Beacon Press).

Markus, T.A., 1993, *Buildings and Power*, (London: Routledge).

Markus, T.A. and Cameron, D., 2001, *The Words between the Spaces: buildings and language*, (London: Routledge).

McLennan, P. and Cassels, S., 1998, New Ways of Working: freedom is not compulsory. In *Facilities Management: the New Agenda, Proceedings of the BIFM Annual Conference, Queens College Cambridge*, pp. 72–76.

Reich, R.B., 1991, *The Work of Nations: preparing ourselves for the 21st century*, (New York: Simon & Schuster).

Rifkin, J., 2000, *The Age of Access*, (New York: Jeremy P. Tarcher/Putnam).

Rose, N., 1990, *Governing the Soul: the shaping of the private self*, (London: Routledge).

Rose, N., 1999, *The Power of Identity: reframing political thought*, (Cambridge: Cambridge University Press).

Rosenberg, R.S., 1999, The Workplace on the Verge of the 21st Century. *Journal of Business Ethics*, **22**(1), pp. 3–14.

Ross, K., 1996, *Fast Cars, Clean Bodies: Decolonisation and the Reordering of France*, (Cambridge, MA: MIT Press).

Sawhney, M. and Parikh, D., 2001, Where Value Lives in a Networked World. *Harvard Business Review*, January, pp. 79–90.

Senge, P.M., 1997, Communities of Leaders and Learners. *Harvard Business Review*, Sept–Oct, pp. 30–32.

Sennett, R., 1997, The Search for a Place in the World. In *The Architecture of Fear*, edited by Ellin, N., (New York: Princeton Architectural Press).

Sennett, R., 1998, *Corrosion of Character: the personal consequences of work in the new capitalism*, (New York: Norton).

Shields, R., 1999, *Lefebvre, Love & Struggle: spatial dialectics*, (London: Routledge).

Soja, E., 1996, *Thirdspace: journeys to Los Angeles and other real and imagined places*, (Oxford: Blackwell).

Stark, W., 1952, *Jeremy Bentham's Economic Writings*, (London: George Allen & Unwin).

Sundstrom, E., 1986, *Work Places: The Psychology of the Physical Environment in Offices and Factories,* (Cambridge: Cambridge University Press).

Thrift, N., 1997, The Rise of Soft Capitalism. *Cultural Values*, 1(1), pp. 21–57.

CHAPTER 3

Organisational Change, Representations and Facilities

Michael Fenker

3.1 INTRODUCTION

This chapter looks at the role of facilities management within the management of organisational change by first exploring concepts of internal dynamics, notably those based on studies of organisational learning, and then the relevance of a distinction between built environment and representations perceived through this environment. This leads to consideration of the management of representations as an essential contribution to the management of organisational change.

Participatory action research, carried out over a period of five years in a company in France, allowed the exploration of two contradictory phases of a process of change, and observation of how a lack of management of diverging representations impacts on the understanding of goals pursued by the organisation. This analysis helps identify the role that space plays within organisational learning by making the divergence that may exist between representations explicit. Consequently, we are able to outline how space management could reflect and support change, understood as an iterative process of elaboration.

Since the early 1980s, competitive positioning in a turbulent economy, and the need to adapt to constantly varying conditions of survival, have set the management of change at the centre of business strategy. Models of management of change now often emphasise methods that create strategic and operational flexibility.

The concept of flexibility is also widespread in approaches to facilities management, prompted by a concern among practitioners to participate in strategic issues. As far as facilities management is concerned, flexibility designates adaptability in planning and operating facilities. Many writers highlight the fact that flexibility is considered as an important means for facilities management to contribute to the management of organisational change. Becker explores the contribution of diversity in workplace portfolios to the development of organisational ability, the ultimate expression of which would be 'the ability to remove change as an ad hoc disturbance and make it a fundamental condition of organisational behavior' (Becker, 2001, p. 28; see also Becker and Sims, 2001). Others have investigated functional aspects of flexibility and analysed the role of design for the adaptability of a physical setting (see for example Leaman, 1998).

Whilst recognising that the concept of flexibility is widespread in the facilities management debate and in current practice, there is a lack of research into flexibility that focuses on how facilities management contributes to the management of change. Exaggerating the role of flexibility can be prejudicial to the conduct of organisational change, as will be evident from the case study to be presented here. The intention is not to deny the importance of flexibility, but rather to show that it does not constitute, in itself, a sufficient and automatic response to problems that occur in a context of change. Therefore, this chapter also examines other means that are available to facilities managers who wish to support strategic change in organisations. In particular, social dimensions of change management are given greater emphasis than in most flexibility-centred approaches.

In order to view change management, other than through the adaptability of a physical setting, the analysis is divided into two parts. First, the potential for using insights from management researchers who perceive the problem of change to be a specific issue related to the internal dynamics of organisations is examined. Secondly, the real objectives of facilities management activities are questioned by exploring the concept of management of representations, making a distinction between physical environment and space.

However, theory often makes a limited contribution to facilities management action and developing these reflections cannot be an end in itself. Therefore, we consider them as components of a framework for the analysis of participatory action research conducted in an international company located in France. This company began a process of professional development. The strategy to manage space in relation to the process was based in a specific department, at least during the initial phase, using two approaches. The first approach aimed at radical flexibility in the assignment of workstations and the physical adaptability of the settings to evolving needs. The second approach focused on stimulating learning of a new profession through a new practice in taking up place. The action of a new manager, by ignoring the effect of the second approach, hampered the process and generated a dysfunctional department.

The fact that the analysis is a case study of a single company certainly limits its validity. However, the real interest of our study lies in the fact that it explores two contradictory phases of this process and therefore allows us to investigate the relevance of the management of space which takes into account the representations that occupants summon up in their action in relationship with a given physical setting.

3.2 THE FM CONTRIBUTION TO THE MANAGEMENT OF CHANGE

3.2.1 Internal dynamics

Predominantly, organisational change is studied from a process-based perspective. Van de Ven and Poole (1995, p. 510) define a process as 'the progression (i.e. the order and sequence) of events in an organisational entity's existence over time'. Change, one kind of event, is an empirical observation of difference in form, quality, or state over time in organisational entity. The entity may be an individual's job, a work group, an organisational strategy, a program, a product, or the whole organisation.

The concept of internal dynamics of organisations enables a much clearer understanding of the highly open nature of such a process. On the one hand, the concept has allowed the development of alternative positions to those that consider that the sequence of change events is prescribed a priori. On the other hand, the concept has lead to analysis of the uncertainty that influences management, not only as a phenomenon external to the organisational entity, for example the uncertainty which comes from market turbulence, but also from the perspective of the logic of actors and the complexity of factors governing collective action.

Internal dynamics emphasises the uncertainty which impacts on organised action, notably by analysing it with the concept of ignorance in order to explain events (Milliken, 1990; Weick, 1995) and with the concept of ambiguity of aims which are pursued by the organisation (Ouchi, 1980). Thus, internal dynamics has strongly contributed to consideration of the change process in a constructive mode more than in a prescriptive mode with a pre-specified direction. The analysis of the constructive mode by Van de Ven and Poole (1995) highlights that change is moving towards unpredictable directions depending on experiences that appear within the process. This perspective places problems such as attribution of meaning to new events and integration of collective and individual experiences in the organisation's knowledge, at the centre of the management of change.

The concept of organisational learning, important in these approaches, is presented as an iterative process in which change is occurring as actors progress by trial and error in a search of shared maps to reach a collective action (Levitt and March, 1988). More recently, some authors have studied the problem of elaboration, permanence, and memorising of knowledge through organisational practices, and have suggested the importance of the context in which these practices take place. From this point of view, learning is considered not only as a problem of knowledge but also as a problem of social identity.

In other words, one's main problem with learning is to become a practitioner and not to learn a practice. Within this perspective, some authors focus their analysis on the process of elaboration of an identity embedded in relationship created within a 'community of practitioners' (Lave and Wenger, 1991; Wenger, 1998), some others on the role of artefacts and other elements of the physical environment in the elaboration of a social and cultural practice (e.g. Norman, 1988). These studies give a new place to the spatial dimension in organisational studies; unfortunately, it has not been much considered by facilities management research yet.

The fundamentally open nature of a management situation is not only determined, as some approaches suggest, by events resulting from outside the organisation. Several researchers on collective action place the origin of uncertainty at the very core of the organisation. This perspective is argued by Girin (1990), who defines action as being determined by motivations and contexts. Motivations of action comprise the existence of a result to be obtained, 'of a dependence on the logic of constrained rationality according to Simon' and of results in 'negotiated agreements between participants'. The contexts of action are what determines choices which are made, that is, what allows us to give a meaning to an event, an act, a message. From this perspective, action appears as not easily predictable compromises between motivations and context. They build up, not only by adequacy of means pursuing a result, but also by adequacy of various other finalities aimed by participants. The open nature of action appears in an increased free will for participants to an action.

This exploration of internal dynamics enriches the problem of change management with two perspectives. The first one concerns experience as essential element of learning which calls for a deeper analysis of the role of environment in the processes of (re)formation of an identity and, in a corollary way, in the elaboration and transmission of knowledge. The second perspective concerns the increase of free will, which leads to the consideration of change, not as a process controlled from without an action any more, but as a process of elaboration of compromise between participants. The relative importance of the role of flexibility in the management of change can actually be measured from this double perspective. Without reaching a point where we would ask whether this perspective delimits the contribution from facilities management to change management, it raises questions about the actual object of a contribution from facilities management.

3.2.2 Contribution of facilities management to change management

Facilities management can be defined as a process ensuring that buildings and other technical systems support the operations of a company. In spite of the widespread idea that an organisation exists through a double system, a social one and a technical one, the action of facilities management often appears ambivalent: does it deal with the technical system, the physical setting, or the relationships between individual or collective subjects with a physical setting? A kind of autonomisation of the intervention on the technical system to ensure its adaptation to operations provides evidence of this ambivalence. The impression of ambivalence is maintained by an often indistinct use of terms as 'physical setting' and 'space' which refer to different contexts. It seems to us crucial to know whether facilities management considers physical setting or social relationship mobilised through this setting as the object of its contribution to management.

In connection with this, a text by Michel de Certeau (1990, p. 173) proposes an interesting approach by defining space as a meeting of motives. For the author, space is produced by the set of movements that occur within the physical environment. He writes that 'space can be considered as the effect produced by operations which direct it, give it circumstantial account and temporality, and lead it to operate as a polyvalent unit of conflict programs or contractual proximities'.

We find a more profound study of this idea in Lautier' *Ergotopiques* (1999) where he positions the concepts of space and physical setting in relationship with each other. Lautier (1999, p. 176) writes: 'One can speak about space as a system of social relationships, mediated by a physical organisation.' Space therefore refers back to the diversity of positions in social relationships between occupants of a place. For Lautier (1999, p. 214), space 'is the representation which organises the perception of this setting for a (collective or individual) subject'. Thus, space differs according to the representations subjects have, depending on their position, their interest, their origin, or their project with regard to other subjects. Different representations and different points of view on a situation border to each other, superimpose each other, and match each other in 'spatiality', if we give this name to the plurality of spaces. The resulting concepts of conflict and instability of perceptions lead Lautier (1999, p. 67) to underline the insight that 'Managing space is, at first, managing the representation.'

The concepts and meaning which subjects impose on representations, their contradiction and their evolution, should be at the centre of a preoccupation that perceives space as the real object of facilities management. In this case, reflection on representations related to the physical setting is especially important, as they necessarily deal with objectives pursued by a collective or individual subject, or with the way to achieve this objective.

Internal dynamics, whether it refers to an iterative process of building up of meaning or to a greater free will of the participant, constantly generates situations that evaluate the physical setting as the representation that subjects have developed about the action. There is either a correspondence or a disconnection between what the physical setting 'proposes as reading' and what the situation 'requires as reading'. The evolution of situations would then need to manage through time the relationship between action, representation and physical setting, in order to avoid any disconnection. Next, the concept of elaboration of representations, because of its proximity with the concept of elaboration of knowledge through experiences, as analysed by Argyris and Schön (1978), would allow us to consider a contribution of facilities management to the management of change, which, though complementary, would be broader than the search for flexibility. It is then necessary to have a better understanding how representations could be managed. The value of the case study presented in this chapter is to make a step in this direction by highlighting the consequences of a lack of taking representations into account.

3.3 CASE STUDY

The concepts introduced thus far will now be applied in a case study, which allows us to see how the management of representations might play a crucial part.

3.3.1 The company's activities and the problem of extending them

The company is an office furniture manufacturer. Its European headquarters is located in Eastern France. Historically, the company's activity is based on a logic of production and sales of furniture. Confronted with high competition, the company in the mid-1990s decided to extend its activity to the services sector and more particularly, the consulting sector. This choice was more risky and difficult to achieve than choosing to compete via product quality and price.

Indeed, instead of confining itself to a well-known profession, the company needed to evolve its abilities and mobilise new professional and cultural references. The plan to retain the personnel emphasised the extent of difficulty: the evolution of profession does not only need a transfer of a set of techniques and knowledge, it also asks the employees to change their ways of acting and their ways of working.

3.3.2 Participatory action research: the marketing department

The approach developed by the company to achieve this change is the heart of this analysis. Specifically, it looks at the relationship between this approach and the forms of occupation and management of space.

The analysis is based on participatory action research, carried out for about five years in this company. The observations concern mainly the marketing department and its 40 members. This department has a hinge position in the company. On the one hand, it is in charge of a great part of the mission of development of services, and on the other hand, actions of this department have a considerable impact on processes that give structure to the activity of the whole company. Therefore, much of the reflection about the evolution towards a new profession concentrates on this department. It is not surprising then, that it is in this department that the search for new practices leads to experimentation. The form of this experimentation is remarkable, especially as it is strongly based on the design of workplace. This is why the marketing department provides an interesting case study to question relationships between the management of change and the management of facilities.

The collection of information comprises interviews of actors of this process, evaluations performed on behalf of the Directorate and, above all, observations as a member of the project team and also as an occupant of the workplace in question.

3.3.3 The project emerged

Combining proposals for change with the actual setting of the marketing department is made gradually. P, who arrives in spring 1997 as marketing manager, has no intention of using this setting as a learning object for a new profession when he announces his intention to modify the marketing organisation. P wants to increase the relevance of products developed by the department and to 'reinforce the company's position as a contributor to added value and expertise for its customers'. He thinks of creating an organisation within which sub-departments are not longer organised according to the product type, but according to the 'logic of applications' of client products. At this point, the concept of evolution of profession is not central to the changes being considered for the marketing department. As a consequence of this announcement, the argument for a banal project of re-implantation of the department on the same site (that had started before P's arrival) is crushed.

In the following stage, P's department is used as a pilot project for the transfer of the whole company to another site in the region. This project becomes the first achievement of an operation that will last several years. Re-positioning the marketing department building project enables P to give a better visibility to his reorganisation project. Space is above all given the role of tracing an organisational change via a location change. The problem of evolution of profession is not yet present in P's interest in the building project.

The opportunity to combine both problems becomes evident when P starts to define the objectives of the reorganisation more precisely. He wants to increase the skill of each team and to improve flexibility in the allocation of human resources to the marketing process. In order to achieve this goal, P wants to improve knowledge of the marketing processes and to make the transfer of this knowledge easier. This coincides with a meeting between P and a group of experts in workplace design belonging to the marketing department. The build up of a project team, including the author, gives the opportunity for reflection about the impact of these intentions on the design of the physical settings. This reflection emerges from the analysis P makes of the marketing process, with the objective to improve the legibility of the different steps. This work results in an identification of three main steps with their corresponding specific expertise. The first step deals with the development of knowledge specific to the company's profession and with a market analysis; the second step deals with portfolio management and with the development of new products; the third one deals with bringing to the market what is 'produced' by the marketing department and includes communication activities.

3.3.4 Planning within three poles

The objectives of helping the department's personnel to recognise these steps and of promoting identification with the expertise specific to each step, leads the project team to decide to arrange people on the floor plan according to their participation in one of these steps. Characteristics of activities appropriate to each step are analysed by the team to guide the design of a specific setting for three distinct zones called poles:

(1) A first pole is designed for 'experts' activities (eight persons). They collect and analyse information from without the company and bring their expertise to projects. They spend 30 per cent of their time on site. Their main activities are to concentrate and to exchange rather in a non-programmed way. The setting of this pole comprises four individual and non-assigned workstations, consultation and exchange workstations, and individual storage units.

(2) A second pole is designed for 'integrators' activities (twenty persons). They have marketing know-how, co-ordinate and give structure to information. They spend 60 per cent of their time on site. Their activities are to concentrate and to exchange within project teams. The setting of this pole comprises small individual and assigned workstations and a large opened and bordering zone for the collective work of these teams. This latter zone can be transformed by the users to meet specific needs, for example for the purpose of a project or for a simple meeting.

(3) A third pole is designed for 'implementers' activities (ten persons). They produce marketing tools and supports. They spend 90 per cent of their time on site. Their main activities are to formalise and to exchange. They are provided with assigned individual workstations and additional tables to receive visitors while remaining at their workstation.

Organising persons within the premises according to their participation in one of the three steps does not change the fact that they belong to a functional team (sub-department). Therefore, each team is distributed among several poles on a floor plan of 450 square metres.

A set of workstations is provided to each pole, designed for activities appropriate to the process step involved. Extra workstations for activities independent or complementary to these three steps are organised in a fourth zone: additional meeting rooms, concentration cells, a relaxation zone, a multimedia and videoconference room, and a documentation centre. These places can be accessed by all the users of the marketing department, by persons from other departments as well as by visitors.

This is an activity-based setting, where people change places depending on the tasks they perform. This principle is applied to all the four zones. Mobility of persons, which is not limited to this site only, is promoted by remote communication technology equipment. In the fourth zone, a large proportion of these workstations (70 per cent of the whole surface) are not assigned to an individual but to everyone belonging to a pole, or to the whole marketing department. The principle of individual allocation differs according to the main operating mode in each pole: importance of working in groups and frequency of attendance on site. The reduction of the individually assigned surface entailed a diversification of settings and a reduction of about 25 per cent of the total surface occupied by the department.

Workstations within poles are located in open or semi-open space. The workstations that are acoustically isolated, either for an individual activity or for a collective activity, are located in the fourth zone and are available on reservation. Besides this, there are several open places, designed for shorter or non-planned meetings. They are easily accessible as they are arranged along the circulation that links the four zones together. This circulation, without partitioning, is located at the centre of the floor plan and widens out at several locations, notably around an atrium in order to promote meetings.

3.3.5 Flexibility, legibility and management communication

Under P's influence, the company considers the contribution of facilities management to the pursuit of the marketing departments objectives mainly by the implementation of two principles: on the one hand, the presence of several forms of flexibility and, on the other hand, what is specific to each process step in order to promote the understanding of practices inherent in the new profession.

Several forms of flexibility have been anticipated in the design of the premises. The first form is the flexibility in the assignment of places. The number of shared workstations is such that the floor plan can accommodate an additional one fifth of the number of occupants without any modification of the physical setting. Such a form of flexibility has several advantages: (1) to absorb fluctuations in the number of people working on the floor plan, (2) to be able to change the allocation of persons to projects without necessarily changing their location on the floor plan, and (3) to provide a greater flexibility in the assignment of places to projects.

A second form is flexibility in the configuration of settings. It is enabled by the use of construction techniques like the standardisation of dimensions of furniture elements and space partitions, the independence between space partition and other fitting-out elements, flexible connections between workstations and cabling channels, also wireless connections for communication devices. A less important but immediately accessible reconfiguration is made possible due to the mobility of some furniture elements, mainly tables, pedestals and panels.

The other principle implemented in these settings concerns the legibility of the specificity of activities of each process step. A first step in this direction deals with the distinction between the three process steps by gathering activities in three poles. Each pole is provided with a different setting. The name given to some elements of the setting gives evidence, either of the strong presence of these elements in the occupants' practice, or of a kind of judgement. Different high and low 'consultation' tables and 'reflection' workstations are available for experts, while integrators have a large re-configurable 'project room' for different kinds of meetings and a small 'box' for individual work, and implementers are using individual workstations equipped with big personal computers which have, in this zone, a strong visual impact. The design of each pole creates an arrangement which designates, through the attendance and movement of occupants, different ways of working. The expected benefit from this design is to contribute to identification with ways of working within each pole, and to establish an identity for each professional group, this identity being able to support 'values' that the new profession represents for the company.

Apart from these two principles directly related to the design of the premises, P perceives the necessity, for a successful change, to convince the members of the marketing department of the validity of the approach by explaining the motivations, the design and the expected results. Communication about the approach starts at the moment of the elaboration of the design concept. The communication content keeps on getting deeper throughout the realisation of the project and continues to evolve with occupation of the setting.

Communication is performed through several channels: written documents (memos, welcoming letter when moving into the new site), work and explanation meetings about the project, meetings to evaluate the difficulties in the realisation of change during the weeks following the moving in, and with the actual attendance of P on the premises.

In fact, through a demonstration of the use of the premises, P displays a working style that 'conforms' to the representation of an 'efficient way of working' which he had himself previously built and diffused. Thus, it is important for him to display a working style that encourages collaboration and exchange between people in different places. He also makes full use of the diversity of spaces for his individual activities.

The planning principles, as well as the communication about the approach, later prove to have an important contribution for the elaboration of representations.

A first evaluation performed one year after moving in supports the management's intentions. A series of dysfunctions are recorded (technology, room accessibility and concentration problems) but users generally confirm their acceptance of the setting, with a relatively high global satisfaction, the perception of an efficiency of the physical setting, and the value given to the experimental character. It can also be noted that the absence of traditional symbols of different hierarchical levels in the setting is highly appreciated by the occupants.

3.3.6 A change in management

The purchase of the company by an American group coincides with the arrival of the manager D, successor of P, two years after moving into the new site. This change at the head of the marketing department corresponds to discontinuation of the operations previously implemented. It is not the purchase itself that calls implementation of the profession evolution into question, as both companies are of the same opinion on this matter. The reasons that lead D to decide to modify the department operation, together with its spatial dimension, are to be found elsewhere. Two hypotheses can be called up to explain his position.

The first hypothesis links with the argument made by D of an operational maladjustment causing short-term difficulties in product development. D notes a weakness in the co-operation within the sub-departments, exaggerated by a dispersal of department members and too frequent journeys. D comes back to the idea of directing the process more towards a 'sub-department activity' approach again.

The second hypothesis is suggested by the perceptions of other members of the marketing department. They take into account a management mode by D, which cannot be easily reconciled with the physical setting implemented. The number of formal management meetings, the holding of project meetings to which D attends in closed rooms, as well as the importance given to the formalisation of new operating rules are some examples of an operation which comes up against the setting or against the occupants' perception of the setting.

Given the lack of knowledge about the real motivation behind D's position, some changes are observed, directly related to the physical setting, quickly occurring on the field. The request of a greater attendance of people on site which begins to have some effect and the realisation of a closed office assigned to D are the most immediate changes.

Because of these changes, increasing confusion among the department members is perceived. Many people come up against the newly introduced modifications because of their fondness for the previous operation and setting, as revealed by an evaluation performed few months after D's arrival. The realisation of D's office is an example of it. Before anything else, people associate the permanent assignment of this office with D's hierarchical position. Therefore, this office is a negation of the previous principles of allocation. The subsequent confusion has some effects, which notably show themselves in co-ordination difficulties for some activities.

Undoubtedly, D ignores the identification of a great number of persons with the initial setting when he orders a wider transformation of the premises. D also ignores that he is dealing with an organisational and spatial setting that has been built through time and should be cautiously managed in order to take advantage of the effects it produces.

In the context of the transformation project, D asks the project team, who had already been involved in the initial planning, for an evaluation study. Even though this evaluation eventually did not have a great impact on the re-setting, it is however very useful to analyse the discontinuity that occurred.

3.3.7 An evaluation of the perception of change

The objective of this evaluation was to understand the efficiency of the management style and its possible relationship with the setting. On the one hand, this involved evaluating the perceptions of factors which would play a role in work efficiency amongst members of the marketing department, and, on the other hand, to record the diversity of comments and assess the coherence or divergence of opinions. A qualitative approach was chosen and information was collected by means of semi-structured individual interviews performed with one third of the personnel. Interviews were conducted from themes that were identified in relationship with their importance in the initial setting or in the operation initiated by D, such as privacy, interaction, concentration, accessibility. The syntheses of these interviews is then presented and discussed by all the department members, and included in a report given to D.

Analysis of the interviews revealed that most people perceive a relationship between the management style, practices within the department, and the work environment. Those interviewed generally stated that this relationship undergoes transformations that are not positively perceived. The lack of coherence and continuity was most deeply felt, as illustrated by this statement: 'we are independent professionals, but too obviously dependent'.

The evaluation also reveals the diversity of representations that governs the perception of the efficiency of this environment. This is illustrated by the fact that there are as many divergent versions as interviews to describe the governing factors for situations of concentration, interaction, management, etc. Interviews reveal, not only the diversity of situations faced by actors, but also the paradoxes which may exist within a given situation, as noted by the following statement: 'we want control and we want direction'.

The identification with previous practices expresses most often itself through the fondness for the concept of activity pole. People recognise themselves in the roles of 'expert', 'integrator', and 'implementer'. A great value is perceived with the existence of poles as they provide people with a reference mark for activities and give a feeling of cohesion.

3.4 DISCUSSION

Considering the two phases that observation has identified during a specific time in the life of this company, the differences between them are striking. The aim is not to return to the differences between management styles, as this is closely related to P's and D's individualities, nor to the organisational choices themselves as they are difficult to compare insofar as the objectives were different. The interest is in understanding the different approaches that have been developed to manage the pursuit of objectives in relation to a physical setting. Within these approaches, two conceptions of the contribution of the management of space to the management of change emerge with different effects on the progress of the process. Analysis of these issues provides us with valuable indications of the role of facilities management in the management of change.

P's approach relies, on the one hand, on flexibility, and on the other hand, on the understanding of the work process by the department members and on the development of an identity within the new practice. D's approach relies only on flexibility to introduce another operation, notably when he asks for an increased attendance of the marketing department members, or when he has his own office installed on the floor plan. His management of the premises has essentially a technical nature. This fragmentary approach has created a vicious circle where initial practices and the space on which they rely are called into doubt again by a new operation, such as when the new operation is questioned by the presence of a space elaborated through the previous practices. The department dysfunction is therefore produced by a situation where a new practice that leads to a new representation comes up against an existing representation without the shift between these both representations being managed. Thus, to manage representations amounts to linking previous and new practices.

The importance of the physical setting reveals itself in the understanding of a game of relationships where practice, as it takes place into a setting, produces or re-produces a representation, and where this representation, as it is mobilised in this setting, links to it. When we say that a subject mobilises a representation through a setting, this representation is of course not intrinsically contained by this setting. It only exists on condition that practice keeps on bearing it, reproducing it. But practice cannot take place without a place.

The management of representations is thus not only conducted by the management of facilities. It principally takes place during an action between participants elaborating a compromise. This elaboration covers the shift between a set of different representations mobilised in the action. This shift calls for a setting modification only when it becomes too large.

Let us recall that representations cannot be decided. Some of them are mobilised by an individual subject while others are mobilised by a collective one; some representations are evolving quickly while others last a longer time. Observations performed on our precise field have shown that the relationship between a representation concerning the organisation and a representation concerning the physical setting has been built through time. Besides, this process did not stop at the moment when the setting was made available to its occupants.

3.5 CONCLUSIONS

To manage representations in a company needs an understanding of the process of evolution of representations of work or of the organisation, and how and when representations of the physical setting are not relevant any more. The management of this permanent unbalance calls into question sometimes the representation, sometimes the setting. Such a conception of management of space could transform the design project and its management through time into an object of production and a sharing of knowledge on the objectives pursued by the organisation.

This study has shown the importance for facilities management of understanding the representations mobilised by occupants during their action. It has revealed the prejudice that a lack of articulation between different representations and between representation and environment may cause to an organisation. The concept of managing space as an articulation between representations and a built environment could be an essential contribution to the management of change and to the management of the problem inherent in it, notably organisational learning. Nevertheless, this perspective is largely dependent upon the contribution of a greater number of researchers in order to study the questions that are raised about the operational nature of such an approach. However, the interest in the spatial dimension shown by some streams of management science, for example by examining the 'contexts', the 'situated action' and 'artefacts', points out the opportunity and the relevance for joint research which would fully integrate the knowledge acquired in the field of facilities management.

The duration of our involvement and the form of case study observation emphasises the constraints and the difficulty of developing a thorough knowledge of relationships between representations and environment that are built up within an organisation. Beyond problems that are then faced for extensive researches of this type, it seems necessary to question the level of integration of the competence to manage representations within the organisation. The concept of ongoing processes and the strategic dimension of the management of representations should fuel the debate about the limits of an outsourcing of facilities management competencies.

3.6 REFERENCES

Argyris, C. and Schön, D., 1978, *Organizational learning: a theory of action perspective*, (Reading: Addison Wesley).

Becker, F., 2001, Organizational agility and the knowledge infrastructure. *Journal of Corporate Real Estate*, **3**(1), pp. 28–37.

Becker, F. and Sims, W., 2001, Offices that work: balancing communication, flexibility and cost, Cornell University, International Workplace Study Program. Available http://iwsp.human.cornell.edu/pubs/pdf/IWS_0002.PDF (accessed 6 May 2003).

Certeau, M. de, 1990, *L'invention du quotidien, Tome I: Arts de faire*, (Paris: Gallimard).

Girin, J., 1990, L'analyse empirique des situations de gestion: éléments de théorie et de méthode. In *Epistémologies et sciences de gestion*, edited by Martinet *et al.*, (Paris: Economica), pp. 41–182.

Lautier, F., 1999, *Ergotopiques: Sur les espaces des lieux de travail*, (Toulouse: Octarès).

Lave, J. and Wenger, E., 1991, *Situated learning: legitimate peripheral participation*, (Cambridge: Cambridge University Press).

Leaman, A., Bordass, B. and Cassels, S., 1998, *Flexibility and adaptability in buildings: the killer variables*, (London: Usable Buildings). Available http://www.usablebuildings.co.uk/ (accessed 6 May 2003).

Levitt, B. and March, J., 1988, Organizational Learning. *Annual Review of Sociology*, **14**, pp. 319–340.

Milliken, F.J., 1990, Perceiving and interpreting environmental change: an examination of college administrators' interpretation of changing demographics. *Academy of Management Journal*, **33**(1), pp. 42–63.

Norman, D.A., 1988, *The psychology of everyday things*, (New York: Basic Books).

Ouchi, W.G., 1980, Markets, bureaucracies, and clans. *Administrative Science Quarterly*, **25**, pp. 129–141.

Van de Ven, A.H. and Poole, M.S., 1995, Explaining development and change in organizations. *Academy of Management Review*, **20**(3), pp. 510–540.

Weik, K., 1995, *Sensemaking in organisations*, (Thousand Oaks, CA: Sage).

Wenger, E., 1998, *Communities of Practice: learning, meaning, and identity*, (Cambridge: Cambridge University Press).

CHAPTER 4

Space Use among Growth Companies: Linking the Theories

Jan Bröchner and Paul Dettwiler

4.1 INTRODUCTION

How do growth companies handle their space needs in successive stages of their evolution? Most studies of workspace changes have dealt with existing large companies rather than with small and growing companies. As publications extolling the 'new economy' multiplied, few people seem to have noticed that the eager converts who espoused new office layouts for networking tended to be large, old and mature companies (see e.g. Rifkin, 2000). A dominant theme has been how these companies can transform their facilities so that changes in work practices brought about by networked personal computers, web tools and mobile telecommunications can be accommodated and exploited. It is rare that there is an analysis of how small and growing firms manage their needs for space; when O'Mara (1999) discusses incrementalism in this context, it is an exception.

Apart from the majority of investigations, formed by case studies, there are almost no broad surveys of space use, and these indicate a per capita space reduction when companies grow. A study of 90 firms in four European cities in the early 1990s found that an increase of employment over time by one per cent reduces the space consumption per office worker by about 0.2 per cent, which is interpreted as an indication that firms face adjustment costs for relocation in the office market (Romijn et al., 1996). Moreover, surveys of large UK organizations in the late 1990s (Gibson, 2000; Gibson and Lizieri, 2001) show that there is a slight reduction of work space per employee due to new working practices, but these surveys throw little light on growth firms, except the observation that smaller firms with up to thirty employees may benefit more from serviced and shared conference rooms and similar spaces in offices.

Adding to the lack of broad empirical investigations of space for growing companies is a near absence of conceptual foundations for the study of growth. An exception is the premises ladder model, presented by Blyth and Worthington (2001, p. 43), with four stages of organizational development and their impact on the degree of specialization required from buildings. In this model, originally developed for manufacturing firms, the stages are Infant, Youthful, Mature and Established, and the emphasis is on how facilities show increasing functional specialization for each of these stages. Whether this is a fruitful conceptualization for the study of growing professional services companies and other facilities users outside manufacturing is an open issue. Thus, we need a more general base for understanding how growing companies manage their space needs.

Our purpose here is to relate theories of company growth to theories of space use and relocation. The company size range we consider is in the interval up to about 500 employees. First, we shall provide a short review of main contributions to the theory of growth in phases, with particular emphasis on spatially related resources and features ascribed to individual phases. A second approach is through architectural theories of functional and symbolic uses of facilities, followed by a consideration of how companies may adjust their space or their work practices as they grow. Finally, these theories are linked in a simple, three-phase model where the spatial consequences of company growth are identified[1].

4.2 LIFE CYCLES OF GROWTH COMPANIES

Penrose (1959) saw the growing firm as a collection of productive resources that render services; these resources are physical or human, and they interact. Physical resources are bought, leased or produced for the use of the firm itself. Managerial services are rendered by human resources. She suggested that unused productive services or resources are a selective force in determining the direction of expansion of a firm. Obviously, the facilities used by a firm belong to her category of physical resources. However, there is little or inconclusive evidence of whether facilities and location have an independent effect on the development of growing firms (Deakins, 1999, p. 212). Instead, we shall concentrate on the effects of growth on facilities rather than vice versa.

4.2.1 Growth in phases

While Penrose did not formulate a life cycle model for new and growing firms, others have done so. One inspiration has been general theories of economic growth (Churchill and Lewis, 1983). When Greiner (1998) looked back at his early (1972) contribution on evolution and revolution in growing organizations, he noted the existence of a consensus among researchers that there are major phases of development, although the precise length (3–15 years) and nature of the phases are debated. In his view, each phase is characterized by a typical evolution and ends with a crisis or revolution, as shown in Table 4.1, which refers mainly to industrial and consumer goods companies.

Greiner speculates that a possible Phase 6 would be characterized by extra-organizational solutions like creating a holding company or cross-ownership, alliances and network organizations. In Phase 6 the growth firm would also investigate opportunities for being acquired by another company.

---

[1] The work presented here is part of the 'Interfirm Relations in Facilities Management for Growth Firms' project under the national Swedish research program 'The client with the customer in focus', launched in 2000, involving several universities and supported by the Swedish Research Council for Environment, Agricultural Sciences and Spatial Planning (Formas).

More recent experiences from consulting, law and investment firms led Greiner (1998) to reformulate a four-phases model for growth in professional service firms, as summarized in Table 4.2. Although the life cycle models in Table 4.1 and Table 4.2 differ in several ways, we can see that strategic changes of facilities with relocation are associated with phases 2 and 3.

Table 4.1 Five phases of company growth (based on Greiner, 1972, 1998).

Phase 1	Phase 2	Phase 3	Phase 4	Phase 5
Creativity	**Direction**	**Decentralization**	**Coordination**	**Collaboration**
Creation of a product and a market. Informal and frequent communication. Increased numbers of employees. Ends with *crisis of leadership*	New management, specialization (manufacturing-marketing), hierarchy grows. Dissatisfaction among low-level managers. Ends with a *crisis of autonomy*	Delegation. Product group(s) launched. Infrequent communication from top. Acquisition of outside enterprises. *Crisis of control*	Formal planning, decentralized units are merged into product groups, Too large organizations to manage. Major organizational changes. Lack of confidence. *Red tape crisis*	Matrix structure for senior management. Cross functional teams, consulting staff experts. Experiments with new practices. Self-discipline and social control. *Internal solution crisis*

Table 4.2 Four phases of growth in professional service firms (based on Greiner, 1998).

Phase 1	Phase 2	Phase 3	Phase 4
Entrepreneurial	**Focus on one service**	**Expansion**	**Institutionalization**
The firm tests a variety of market paths. Partners argue whether to stay together, concentration on one vision for the future.	One major service is selected as a focus. Opening another office, adding additional services?	Geographic and service expansion. Ownership struggle between old and new shareholders who bring in new clients.	Institutionalizing the firm's name, reputation and standard way of operating. *Crisis of cultural conformity.*

Churchill and Lewis (1983) identified five stages of small business growth: (1) existence, (2) survival, (3) success, (4) take-off, and (5) resource maturity. For each of these five stages, they highlighted the relative importance of key management factors: four company-related factors (financial, personnel, systems, and business resources) and four owner-related factors (goals; operational, managerial and strategic abilities). In the context of facilities management, the facilities would constitute a business resource, while the ability to manage facility needs would form a subset of operational and managerial abilities.

Deploring the lack of a theoretical base in most of the early contributions to life cycle models, Garnsey (1998) has taken her starting point in Penrose's resource-based view of the firm and concentrates on early growth phases. From a facilities viewpoint, it is worth noting that Garnsey emphasizes the special case of firms that start growing in an incubator environment (Table 4.3).

Table 4.3 Early growth phases of firms. From E. Garnsey, A theory of the early growth of the firm. *Industrial and Corporate Change* 1998, 7(3), 523–556. Reproduced by permission of Oxford University Press.

Phase 1	Phase 2	Phase 3	Phase 4	Phase 5	Phase 6
Access resources	**Resource mobilization**	**Resource generation**	**Growth reinforcement**	**Growth reversal**	**Accumulation of resources**
Founders provide assets and impetus. Search activities and perception of opportunities dominate. Use of as far untraded resources. Influence of face-to-face contacts. Raising funds.	A business plan, achieving credibility. Trial and error lead to procedures. Specialist roles emerge. Creating links to founders, customers and suppliers by self-presentation.	Continuity in key relations to customers and distributors. Routinization of problem solving. Self-sustainment, build-up of competence. Reaching minimum size for efficiency.	Growth and further investment mutually reinforcing. Underused staff, as resource generation is organized better. Founding entrepreneurs may find delegation difficult.	Complexity and fluctuations need professional management. Changes in the industrial environment. Upgrading, product renewal or extension of product range. Possible sale of the company.	Market extension through alliances and acquisitions. Early identity of firm transformed.

We thus find that theorists concerned with growth companies distinguish between phases and that a major change is thought to be when professional managers replace entrepreneurs. These writers have associated particular phases with phenomena that can be related to facilities as a resource for the firm. Facilities may support work directly—or indirectly—by emitting signals to employees, customers, suppliers and others outside the company. It also seems that professional services differ from other sectors and that firms that start in incubators are different.

4.2.2 Measuring growth

The growth of a firm is usually measured in terms of changes in its annual turnover or in the number of employees. Although changes in turnover could be a relevant indicator of space needs for firms that carry physical inventories of some size, it is probable that the number of employees is more strongly related to the development of facilities. However, the tendency to work at a distance from company facilities, as well as the presence of both customers and external service providers imply that a linear relationship between employees and space can hardly be expected.

4.3 FUNCTIONAL AND SYMBOLIC USE OF SPACE

We have now identified typical growth patterns for firms and also their varying needs for using tangibles such as facilities to signal, internally and externally, their intangible qualities. Among the wealth of architectural and space use theories, there is thus reason to look for approaches that acknowledge the interaction between basic space needs and space used as signals. Inspired by Gregor Paulsson, who had distinguished between the physical and symbolic aspects of the environment, Norberg-Schulz in his Intentions in Architecture (1963) identified four dimensions of the building task of architecture: physical control, functional frame, social milieu and cultural symbolization. Many other writers on architectural theory have proposed similar classifications. Here we shall restrict ourselves to only two categories, namely functional use and symbolic use of space.

4.3.1 Functional use

A clear expression of the functional frame mentality, which assumes that space requirements should be derived from human worktasks is the Neufert (2000, first German edition in 1936) approach. Precise quantitative measures are typical of normative functional frame thinking. Taking offices for an example, work practices in the office are assumed to be closely linked to the available technology. A functional influence on space use for 'new economy companies' is that they are believed to accumulate relatively little paper and require less space for storage; also, the nature of their work suggests closer collaboration between employees (Latshaw *et al.*, 2001). Thinking in functional terms, both these phenomena should decrease the total space need per employee, in addition to the effects of working away from the office: hot-desking, teleworking. The functional frame mentality is still operative when we design workplaces intended to support organizational learning and networking, although the analysis has to start with an understanding of work processes, old and new.

The deterministic, functional minimization or Neufert approach to office space needs and office design does not seem an obvious choice for companies that grow, especially when there is uncertainty about future work tasks, task volumes and supporting office technologies. It also plays down architectural features that constitute cultural symbolization.

A particular development of functional thinking can be read into the question posed by Alexander *et al.* (1977, p. 690): 'Is it possible to create a kind of space which is specifically tuned to the needs of people working, and yet capable of an infinite number of various arrangements and combinations within it?' They pointed out that there was more genuine flexibility in large old converted dwellings containing spaces that support work groups; although walls cannot be moved, the house is adaptable. However, their discussion illustrates a tendency, still present in the late 1970s, to downplay the signals that radiate from operating a business in a lavish Victorian residential setting. Today, it seems that the functional and the symbolic uses of space are brought closely together in such a context.

4.3.2 Symbolic use

Gilbert Scott in his *The Architecture of Humanism* (1924) begins with the Vitruvian trinity of commodity, firmness and delight, but soon reaches the point where he proclaims the gravest fallacy of Romanticism to be that it regards architecture as symbolic. His argument against the study of associative values is that these values 'will be determined wholly by the accidents of our time and personality'. Nevertheless, it is precisely the accidental nature of architectural signals that could explain their importance for growth companies. The volatility of symbols has to be recognized as a problem, however. 'Say Good-Bye to Grand Granite Interiors' is the heading chosen by Zelinsky (1997) for her discussion of how the palatial corporate settings of the 1980s became a symbol burden. In itself, ownership of real estate has a symbolic value, though nowadays weakened because of fundamental and global economic changes implying that access to space and services count for more (Rifkin, 2000).

Writers on corporate culture sometimes address architectural signals. Berg and Kreiner (1990) discuss how buildings have become increasingly important in expressing corporate identity. Thus, buildings may contribute to the symbolic conditioning of organizational behaviour, act as totems, symbolize a strategic profile, function as packaging, or symbolize status, potency and good taste. They can also be interpreted as markers of time, ideas and existence. Evette (1993) describes how corporate architecture may express managerial attitudes to supervision and communication.

A firm, especially in professional services, that wishes to project the image that it is creative, agile and virtual may choose to design its offices so that they include references to homes, TV studios, bars and cafés, an architectural eclecticism that picks style elements from several building types with increasingly differentiated designs during the twentieth century. In the last decade, symbolic design and interior decoration that supports an image of being a learning organization has included style references to the declining sector of manufacturing by an allusive use of metal and glass in the office environment. Furthermore, if this has been in the setting of a converted plant or warehouse, the contrast between the origin of the building and the activities of the growth firm is even greater and constantly signals to both employees and visitors a considerable mental distance from the traditional manufacturing firm. This development of ironic features can be attributed also to an obsession with a playful style of work, intended to attract and retain staff (van Meel and Vos, 2001) or linked to the theatre analogy that permeates current ways of thinking about companies, most clearly manifested by frequent recourse to the actor metaphor.

4.3.3 Are functional and symbolic features separable?

Davis (1984) in his overview of earlier research on influences of the physical environment in offices distinguished between physical structure, physical stimuli and symbolic artefacts in the environment; he also suggested that these elements are best viewed as alternative ways of construing physical phenomena. The two central problems in architectural theory are identified by Hillier (1996) as the 'form function problem' and the 'form meaning problem'; buildings viewed as 'social objects' can be thought of as combining function and meaning. It is unlikely that we shall be able to, or even wish to, separate the functional and the symbolic entirely, especially when the symbolic element is an exaggerated expression of a (new) functional feature of office design. Professional services companies are found with physical layouts that allow, or even force, people to interact with one another (Hargadon and Sutton, 2000).

4.4 SPACE ADJUSTMENTS

Most growing companies have several alternatives to choose between when the number of employees goes up. Space needs can sometimes be reduced by a change in work practices. Within the existing space, it is often possible to reconfigure the layout so that occupancy is intensified by denser seating arrangements. An increase in the available space for the firm can be brought about by expanding in the same geographical location, at the same address, or more often by partial or total relocation of activities. Mergers and acquisitions often have effects that amount to at least partial relocation.

When a growth firm reaches maturity, we should expect that managerial abilities have been extended to a professional treatment of facilities provision and use. It is only now that we can expect a focus on a conscious strategy for reducing facility-related costs and refining workplace design (cf. Becker and Steele, 1995; Vischer, 1995). It is also in this stage that concepts such as Process Architecture (Horgen *et al.,* 1999) acquire relevance, since the effectiveness of design tools such as collaborative workshops where staff are able to interact with professional space managers will be more evident.

4.4.1 Reducing space needs

In a situation where the growth firm has what is perceived as excess space, a reduction of space needs per employee can be achieved in more than one way. There is the possibility of having employees increase the proportion of their time that they spend away from the office, primarily by greater reliance on teleworking and other flexible work practices (Avery and Zabel, 2001) but also by working in transit or with customers or suppliers at their premises. To some extent, growth firms in sectors that carry inventories of goods may reduce space needs by shifting these to suppliers or consumers.

When employees are present and do not work elsewhere, space is used by them and also by visitors. Visitors can be thought of as customers, suppliers and investors. Suppliers are different because some of them spend long and continuous periods in the facilities, while others come and go more rapidly or never even set their feet there. An alternative distinction would then be between short-term visitors and long-term visitors. The proportion between these visitors is expected to influence the facilities needs. Also here, within limits the intensity of visitor use of facilities can be regulated in order to avoid a need for space adjustment. The choice of museum fee levels is one example.

4.4.2 Reconfiguration

The mismatch is always there between what a building can offer and what the organization needs (Blakstad, 2001). Flexibility and adaptability of the building can raise the ability to reconfigure. Building design often inhibits change in the same location because of complex interaction between the systems of structure, enclosure, services and interior finish, as revealed by analysis of actual renovation projects (Slaughter, 2001).

4.4.3 Relocation

Relocation is costly in itself, directly and also indirectly because of its disruptive effects on work in the firm (Carter, 1999). Many factors can be assumed to affect the decision to relocate (Nourse, 1992; McDougall, 1993; Dent and White, 1998). The distance to suppliers, customers, employees, both potential employees and those who are already there, and competitors, in addition to the cost of using the facility itself, may influence the choice of a new location, as well as the symbolic features of the building and its address. Moreover, the set of service suppliers may change along with location: the degree to which activities are carried out by the firm's own employees or by external service providers can be reconsidered.

Since a growth path can be hard to predict, the decision to stay or to relocate can be affected by the rigidity or flexibility of tenure, the contractual conditions and duration offered for space in a particular location. For the company faced with alternatives of more or less comprehensive bundles of business support services attached to the building, a range from the full bundle typical of an incubator for new businesses to the lease of a naked structure, the clauses and periods in a lease contract may carry great influence (cf. O'Roarty, 2001).

The inability to achieve a satisfactory solution in terms of functional space or to handle the symbolic aspects of a particular setting is thus only one reason for changing the location of the activities of a growth company. Sometimes relocation is needed because a lease expires and the landlord is unwilling to prolong the relationship; sometimes, local authorities make planning decisions that provoke a move. Mergers and acquisitions may create the opportunity or the necessity to relocate. Growth of a highly specialized company may necessitate a wide expansion of the customer base and the decision to work from several sites, perhaps spread over more than one continent.

Many growth firms start in simple accommodation as tenants and assemble their own bundles of facilities-related services. Others spend their initial phase in an incubator, a business park or a technology park where bundles of support services are combined with physical accommodation. These services are sometimes paid for through the rent and may include strategic and marketing advice and other business services such as accounting. When or if a growing firm entirely or partially leaves an incubator or park, it has to rearrange for the support services it needs. It appears to be unusual that firms relocate from one incubator or park to another.

4.5 SPACE NEEDS IN SUCCESSIVE PHASES

Turning back to the Greiner (1972, 1998) five-phase model, there are three observations to be made. First, there is a consensus that there are major phases of development although the precise length, ranging from three to fifteen years, and nature of phases are debated. Here, we choose to distinguish between three phases: entrepreneurial, managerial and consolidation. Any of the more complicated multi-stage models should be viewed with suspicion since they do not take into account the recent impact of Internet technologies and may thus be at least partly obsolete. Second, transitions between the phases are thought to be abrupt; also, the alternatives to growth of the firm are to falter, go into plateau phase, fail or be acquired. We should expect abrupt change to be linked to relocation more often than not, since moving from one location to another is costly. Finally, management may find it difficult to realize that their creation of the organizational setting will later cause a revolution. Since the physical facilities are part of the setting, the symbolic effects of a particular building and its interior decoration may eventually be found counterproductive.

4.5.1 Entrepreneurial phase

In the entrepreneurial phase, the growth company tends to focus strongly on its core activities and on which market paths it shall follow. Assuming that the founders suffer from financial constraints, their facilities policy can be expected to emphasize functional needs rather than spending on symbolic features, which might even do harm to their relation to investors. However, spartan facilities do carry a symbolic message in themselves. Start-ups linked to university research might opt for an informal setting that signals their origin and reflects typical research facilities. Start-ups spun off from large corporations might inherit spatial designs, or they might choose to express a contrast with their original environment. In this early phase, reliance on frequent, informal and face-to-face contacts implies that occupational density should be high. However, uncertainty regarding the company future—and a lack of access to professional knowledge of space management—may cause overcrowding and badly chosen timing of relocation.

4.5.2 Managerial phase

When the growth company enters the managerial phase, professional management has largely taken the place of founders. Focus is still on developing the core business. Workspace layouts can now be assumed to start reflecting hierarchical tendencies when staff roles are increasingly specialized and defined. New problems occur in developing the business, requiring specialists who may express special space requirements. Geographical spread with multiple sites complicate the provision with facilities and facilities-related services. Since management has its focus on the core, issues regarding facilities and support services are dealt with only intermittently. Support costs and the gap between facilities and organizational practice may easily be thought to increase during this phase.

4.5.3 Consolidation phase

The consolidation phase implies continued growth at a steady rate or a plateau. Management is now more efficient and sophisticated than in the second phase, and the ability to match support to core business needs has improved markedly. Whereas the managerial phase could be associated with a lack of focus on facilities and non-core activities, the consolidation phase has its paradox in a greater rate of internal change. While skills in facilities management are available and used more continuously in this phase, they are often matched by higher rates of churn as competition and technology in the core business are subject to change. Reliance on cross-functional teams and calculated experimentation with teleworking and other alternative work practices in this phase can be expected to imply layout changes. Therefore, we identify two similarities with the entrepreneurial phase: informal and frequent face-to-face communication is more important than in the intervening managerial phase, and dwellings can be expected to play a greater role as a workplace for staff in both the entrepreneurial and the consolidation phases. Moreover, exploiting the opportunities for introducing uniform facilities management procedures at multiple sites is another feature that should be found in the consolidation phase.

4.6 CONCLUSION

This paper has woven together concepts from two fields of theory, namely company growth theories and space theories related to architecture and facilities management. We have found that theories of growing companies agree on the existence of several growth phases that can be distinguished. From the viewpoint of facilities management, it appears that a three-phase model is sufficient and fruitful, and that it allows us to predict typical developments in the relation between the organization and its facilities. To understand this relation and how it develops, we also need at least the simple dichotomy of functional and symbolic aspects of facilities, although these aspects tend to be integrated in the workplace.

The transition from entrepreneurial governance to managerial governance of the growing company has been identified here as a critical point with significant consequences for facilities. Companies with a development path dominated by mergers and acquisitions can be thought to have their facilities management also dominated by relocation rather than by slow and continuous adjustments of space.

Results are to be used for empirical surveys of how growth companies organize their facilities and facilities management, and several questions can now be formulated for further research. Do all growing companies conform to a similar pattern of growth, or are there a few basic patterns? Is it true that professional service firms follow a distinctly different path of growth? Is it also true that growth paths have been strongly affected by the sudden advances in telecommunications and web technologies in the mid-1990s? If growth patterns consist of phases with discontinuities ('revolutions') when one phase follows another, will relocation be associated with discontinuities, rather than occur during particular phases? Which is the relation between spatial change and the context of facilities-related services supply? Under which circumstances do growth companies tend to rely on professional FM advisory services, and how?

4.7 REFERENCES

Alexander, C., Ishikawa, S. and Silverstein, M., 1977, *A pattern language: towns, buildings, construction,* (New York: Oxford University Press).

Avery, C. and Zabel, D., 2001, *The flexible workplace: a sourcebook of information and research,* (Westport, CT: Quorum).

Becker, F. and Steele, F., 1995, *Workplace by Design: Mapping the High-Performance Workspace,* (San Francisco, CA: Jossey-Bass).

Berg, P.O. and Kreiner, K., 1990, Corporate architecture: turning physical settings into symbolic resources. In *Symbols and Artifacts: Views of the Corporate Landscape,* edited by Gagliardi, P., (Berlin: Walter de Gruyter), pp. 41–67.

Blakstad, S.H., 2001, *A strategic approach to adaptability in office buildings.* Dr.Ing. Thesis, (Trondheim: Norwegian University of Science and Technology, Department of Building Technology).

Blyth, A. and Worthington, J., 2001, *Managing the Brief for Better Design,* (London: Spon).

Carter, S., 1999, Relocation or dislocation? Key issues in the specialist management of group moves. *Management Research News,* **22**(5), pp. 22–36.

Churchill, N.C. and Lewis, V.L., 1983, The five stages of small business growth. *Harvard Business Review,* **71**(3), pp. 30–50.

Davis, T.R.V., 1984, The influence of the physical environment in offices. *Academy of Management Review,* **9**(2), pp. 271–283.

Deakins, D., 1999, *Entrepreneurship and Small Firms,* 2nd ed., (London: McGraw-Hill).

Dent, P., and White, A., 1998, Corporate real estate: changing office occupier needs—a case study. *Facilities,* **16**(9/10), pp. 262–270.

Evette, T., 1993, Company strategies and architectural demand. In *Appropriate architecture: workplace design in a post-industrial society,* edited by Törnqvist, A. and Ullmark, P., (Göteborg: Chalmers University of Technology, Industrial Planning and Architecture), pp. 53–58.

Garnsey, E., 1998, A theory of the early growth of the firm. *Industrial and Corporate Change,* **7**(3), pp. 523–556.

Gibson, V., 2000, Evaluating office space needs & choices. Report for MWB Business Exchange. The University of Reading, May.

Gibson, V.A. and Lizieri, C.M., 2001, Friction and inertia: business change, corporate real estate portfolios and the U.K. office market. *Journal of Real Estate Research,* **22**(1–2), pp. 59–79.

Greiner, L.E., 1998, Evolution and revolution as organizations grow. *Harvard Business Review,* **76**(3), pp. 55–67 (1972 version in **50**(4), pp. 37–46).

Hargadon, A. and Sutton, R.I., 2000, Building an innovation factory. *Harvard Business Review,* **78**(3), pp. 157–166.

Hillier, B., 1996, *Space is the machine: a configurational theory of architecture,* (Cambridge: Cambridge University Press).

Horgen, T.H., Joroff, M.L., Porter, W.L. and Schön, D.A., 1999, *Excellence by design: transforming workplace and work practice,* (New York: JWiley & Sons).

Latshaw, M., Harmon-Vaughan, B. and Radford, B., 2001, How companies can deliver flexible, effective real estate fast. *Journal of Corporate Real Estate,* **3**(1), pp. 46–55.

McDougall, B., 1993, The changing geography of location. In *The responsible workplace: the redesign of work and offices,* edited by Duffy, F., Laing, A. and Crisp, V., (London: Butterworth Architecture), pp. 112–128.

Neufert, E. and P., 2000, *Architect's Data* [tr. of Bauentwurfslehre], 3rd edn, edited by Baiche, B. and Walliman, N., (Oxford: Blackwell Science).

Norberg-Schulz, C., 1963, *Intentions in Architecture,* (Oslo: Universitetsforlaget).

Nourse, H.O., 1992, Selecting administrative office space. *Journal of Real Estate Research,* **7** (2), pp. 139–145.

O'Mara, M.A., 1999, *Strategy and place: managing corporate real estate and facilities for competitive advantage,* (New York: The Free Press).

O'Roarty, B., 2001, Flexible space solutions: an opportunity for occupiers and investors. *Journal of Corporate Real Estate,* **3**(1), pp. 69–80.

Penrose, E.T., 1959, *The growth of the firm,* (Oxford: Basil Blackwell).

Rifkin, J., 2000, *The Age of Access: How the Shift from Ownership to Access is Transforming Capitalism,* (New York: Tarcher/Putnam).

Romijn, G., Hakfoort, J. and Lie, R., 1996, A Model for the Demand of Office Space per Worker. TI 96–85/5. Tinbergen Institute.

Scott, G., 1924, *The Architecture of Humanism: a study in the history of taste,* 2nd ed., (London: Constable).

Slaughter, E.S., 2001, Design strategies to increase building flexibility. *Building Research and Information,* **29**(3), pp. 208–217.

van Meel, J. and Vos, P., 2001, Funky offices: reflections on office design in the 'new economy'. *Journal of Corporate Real Estate,* **3**(4), pp. 322–334.

Vischer, J.C., 1995, Strategic work-space planning, *Sloan Management Review,* **37** (1 (Fall)), pp. 33–42.

Zelinsky, M., 1997, *New Workplaces for New Workstyles,* (N York: McGraw-Hill).

PART TWO

Innovation and the Innovative Workplace: An Introduction

Brian Atkin

The knowledge economy has already had a powerful impact on the way we work. The productivity of a worker is less frequently measured by how many widgets he or she produces, and increasingly by how many successful ideas he or she conceives and implements. The demands on the knowledge worker reflect this changed emphasis, with increased autonomy, creativity and personal commitment among workers. The movement towards ubiquitous connectivity means that we have further flexibility in choosing our co-workers and work styles. To be successful in the knowledge economy, we need to identify work environments that are conducive to these new ways of working. Moreover, we need to understand and adapt to the demands of sustainable development and construction, if we are to maintain a built environment and workplaces within it that are fit for purpose now and into the future. In this regard, facilities management is concerned not only with ensuring that workplaces function well for their users, but also how workplaces are created in the first instance, how that process is managed and how moves between workplaces may be best accomplished.

The original aim of Part Two was to examine the creation of new workplace concepts, processes, technology and systems to enable flexible working and creativity of the workforce. This has been largely achieved from three different perspectives—one in relation to the creation and delivery of the workplace, another that looks critically at the extent to which workplaces (or rather the buildings that provide them) are adaptable, and from the perspective of sustainability. Kirsten Arge sets the scene in Chapter 5 with her examination of adaptability in office buildings—a topic that has been important in design since the early 1970s. The emphasis has tended to be on physical flexibility and not on functional and financial flexibility. In the 1990s, this changed due to organisational upheaval, new ways of working and use of information and communication technology. Her chapter, *Why Do Real Estate Actors Weight Adaptability Differently for Office Buildings?* examines the motivations, barriers and enablers for providing flexibility in office buildings. The term adaptability is used as the means for assessing the extent to which real estate developers succeed in providing the various measures that will support changes in the workplace into the future. Investing in a high degree of building adaptability is, however, expensive. As we see from this study, actors in the office real estate market are likely to behave quite differently when making decisions concerning adaptability. In particular, actors who build for their core business invest more in adaptability than others.

John Hudson continues the discussion with his reporting in Chapter 6 of the findings of the EC–funded Workspace project. EuroFM was instrumental in initiating the project and was responsible for its management. The project was set up to explore the relationship between the design and management of workspaces and industrial productivity. Six case studies of workspaces in major manufacturing companies are presented in the chapter, *Towards a More Sustainable Industrial Workplace*. Considerable research materials were analysed and from this effort it became apparent that issues beyond the original remit of the project were being raised. The issue of sustainable development appeared particularly important. Hudson's chapter reinterprets the Workspace case studies using a novel framework for understanding sustainable development with respect to four areas—futurity, environment, equity and public participation. The findings are mixed, suggesting that whilst some aspects of sustainability are being addressed, particularly at the local level, there are wider implications that are not.

Lastly in this part, Margaret-Mary Nelson discusses the increasing realisation of the importance of supply chain management as a vital component of research in the facilities management sector. As in a number of other process initiatives being undertaken in this sector, the concepts being tried and tested are management developments borrowed from sectors such as manufacturing, IT and retail. Chapter 7, *The Emergence of Supply Chain Management as a Strategic Facilities Management Tool*, includes a discussion of the findings of research carried out into supply chain management within the facilities management sector. The aim is to explore the concept of supply chain management and what this means in the context of facilities management. Drivers for change and added value in facilities management from the more effective management of the supply chain are examined and some promising areas for further research are proposed.

All three chapters contribute to our understanding of how to innovate in the context of the workplace with the goal of providing space, function and responsiveness to change that can assure the future as well as satisfy the present.

CHAPTER 5

Why Do Real Estate Actors Weight Adaptability Differently for Office Buildings?

Kirsten Arge

5.1 INTRODUCTION

Adaptability in office buildings has been a topic of interest for many years, having evolved from an interest in physical flexibility to one of financial and organisational flexibility. The main focus of this chapter is not what makes an office building adaptable, but how different real estate developers act in the marketplace regarding investments in office buildings where adaptability is a particular concern[1].

Adaptability has long been an important design parameter in the design of some types of buildings, especially housing, offices, hospitals and schools. During the period of architectural structuralism in Scandinavia in the 1970s, three concepts covering different aspects of adaptability were developed: generality, flexibility and elasticity. The three concepts were related to the physical design of buildings and were promoted at the time by some of the leading architects in Scandinavia and by the Swedish Directorate of Public Construction and Property. Later, focus on adaptability waned. There may be several reasons for the loss of interest; for instance, built-in adaptability was too expensive, was not needed or did not function as expected.

During the last five years, however, adaptability has become an important issue once again. In the context of office buildings, the obvious reason is a growing uncertainty about the future and the rate of change in organisations. In the case of hospitals or schools, the basic reasons are the same as for offices, being induced and driven by technology and organisational and/or pedagogical change. In addition, adaptability is now regarded as an important environmental issue. Widespread and costly rebuilding caused by changing functional demands is not in step with a policy of environmental consciousness.

Despite this, office buildings that are currently being planned and built in Norway differ considerably in terms of investments in relation to adaptability. This

---

[1] Ongoing (since 2000) study at the Norwegian Building Research Institute: Adaptability in Office Buildings, Cost and Profitability.

chapter discusses these differences, based on a case study of three types of real estate actor:

1. Private companies that develop real estate to service their core business (which is not real estate development, property rental or facility management).
2. Private or public companies that develop real estate for rental and management.
3. Private companies that develop real estate for sale to investors.

In Norway, private companies that build for themselves, usually own their property as well. Private or public companies whose core business is real estate development, property rental or facility management, own the properties they manage, although they may also sell them. Private companies that develop real estate purely for sale (often referred to as 'hit and run' companies) are, for example, large building contractors, with a subsidiary specialising in real estate development. The investors who buy real estate are private or public funds and they normally outsource real estate and facility management.

5.2 DEFINITIONS

Adaptability has been defined in many ways. In the Scandinavian vocabulary developed in the 1970s, adaptability had three different meanings:

1. generality—the abilities a building has to meet changing functional demands without changing its properties
2. flexibility—the abilities a building has to meet changing functional demands by changing its properties
3. elasticity—the abilities of the building to increase or reduce space according to need.

Gibson (2000) has defined three different kinds of flexibility that closely resemble the Scandinavian concepts of adaptability—see Table 5.1.

Table 5.1 Types of adaptability and flexibility.

Flexibility (Gibson, 2000)	Issues	Scandinavian concepts of adaptability
Physical flexibility	Ease of reconfiguring space Efficiency of space Physical impediments to change	Flexibility
Functional flexibility	Variety of activities/processes undertaken Ability to support movement of staff Ability to change use	Generality
Financial flexibility	Ability to vacate Ability to let/sublet	Elasticity

Blakstad (2001) however makes a distinction between adaptability and flexibility:

- adaptability—creating manœuvring room, space to change, both in the building and in the process, where both have the capacity to answer to unexpected changes
- flexibility—the possibility of changing within a limited set of alternatives, to move and adjust according to a predefined set of possibilities.

In this chapter we nevertheless adopt Gibson's definitions of financial, functional and physical flexibility and, hence, adaptability as the common word to refer to these three types of flexibility.

5.3 CASE STUDY ORGANISATIONS

The organisations used as cases in the study were chosen because they are leading Norwegian organisations in terms of size and position in their respective markets.

5.3.1 Actors who develop real estate for their core business

The four companies in the case study have all moved into new office buildings during the past three years, except the last, Norsk Hydro, whose new headquarters is awaiting the decision to start construction.

Kreditkassen, Norway's second largest financial concern (now sold and part of Nordea), has co-located all its administrative functions and subsidiaries in two interconnected new office buildings called Colosseum Park, housing 1,440 employees. Emphasis has been placed on financial and functional flexibility—any floor in the buildings can be sublet to other companies. The workplace design is that of an open, universal, team-based solution, which permits relocating people and whole sections without disturbing work processes. Nordea has been through extensive organisational changes after moving in, and the open universal solution has proved to be very useful in that regard.

Ergo Group is a publicly-owned limited company producing and selling electronic systems and services, and IT infrastructure services. It has built its own new office building in two phases, due to uncertainties over the size of the company. A total of 900 employees now occupy the building. Emphasis has been put on all three kinds of flexibility. Parts of the building can be sublet to others, because the workplace design is primarily that of an open, universal, team-based solution. Allowance has also been made for erecting walls to provide a 100 per cent cellular solution if required. The building has been developed by, and is owned by, Avantor (see section 5.3.2), but the Ergo Group decided on, and paid for, all flexibility measures.

Telenor is Norway's largest supplier of phones and services to private customers and small companies and this includes mobile telephony and data communication. It has recently co-located all its activities and subsidiairies in new office buildings in several cities in Norway. Earlier, Telenor spent between €875 and €1,500 per employee annually in refurbishing its offices because of organisational and functional changes. Telenor is without doubt the Norwegian organisation that, up to now, has planned for, and invested, the most in financial, functional and physical flexibility in its new office buildings in Bergen and Oslo.

Norsk Hydro is also very conscious about organisational change and flexible work styles. As a leading supplier of oil, gas, fertilisers and aluminium, it is going through large organisational changes at the moment. The new headquarters west of Oslo is being planned to include many of the same flexibility measures as Telenor's office building in Fornebu, Oslo, described in Chapter 1.

5.3.2 Actors who develop real estate for rental and management

Four companies in the case study are among the largest and most professional real estate developers in Oslo.

Entra Eiendom is a recently established limited company, 100 per cent owned by the Norwegian state. Its primary function is to develop or supply central government departments and customers with suitable offices, but other public or private customers are serviced as well. So far, Entra has built only one new office project, which is used as a case here, although it owns and is refurbishing its large, existing building stock. The main tenant in the new office project is the government roads department. The department has been very clear on the workplace design it wants: 50 per cent fixed cellular offices, 30 per cent team-based solutions and 20 per cent cellular offices that can be changed later into team-based solutions if required. The project has been tailor-made to these specifications. The contract is for 20 years.

Avantor is a private limited company with several owners, the largest being Kjell Inge Røkke, who recently took control of Kværner ASA. Røkke brought into the company a large redundant industrial estate, Nydalen, in Oslo, which Avantor has developed into an attractive and modern business area for IT, marketing and media organisations. Nydalen is not among the more expensive office areas in Oslo and probably never will be. Avantor's standard is good, but sober. Its standard in relation to adaptability has developed gradually over time, but it is still rather limited. If customers ask for adaptability, and are willing to pay for extra measures, it will be provided.

Linstow Eiendom is also a private limited company, but with one large owner. It has expanded gradually from a real estate and facility management company into a real estate developer too. Its activities are split equally between the two business areas. Linstow is the owner and manager of Aker Brygge, one of the most fashionable office areas in Oslo, situated on the western waterfront. It has built several office buildings in central and western parts of Oslo. So far, it has had no particular standard with respect to adaptability. Linstow does, however, have a high profile for its architecture, using Norway's best architects to design its office buildings. Recently, Linstow and ROM Eiendomsutvikling (see below) agreed to collaborate on developing the areas owned by Norwegian Railways (NSB) in Bjørvika, a new, large urban development area on Oslo's eastern waterfront adjacent to the Central Station. This area, where Oslo's new Opera House will be built, is expected to develop into a location that will equal Aker Brygge in popularity and cost/revenue.

ROM Eiendomsutvikling is a limited company owned by Norwegian Railways (NSB). Its task is to develop and sell property owned by NSB, and to generate guaranteed profits for real estate investors and its owner. NSB owns real estate all over Norway. In Oslo, Bjørvika is one of its most interesting real estate development areas. Its first office building in Bjørvika will be completed in 2003; it has been designed by the architect Niels Torp, and will be occupied by Scandinavian Airlines System (SAS) and Ernst & Young Consulting and Accounting. In this office building, ROM has introduced most of the adaptability measures needed to give the building sufficient space to accommodate change, as well as a long functional and economic life.

5.3.3 Actors who develop real estate for sale to investors

NCC Eiendom is a high profile real estate development company, owned by a large Scandinavian construction business. It develops office buildings for sale to investors as well as for rental. NCC's most recent development is a high-rise office building at Majorstua in Oslo, in which KPMG is the main tenant. The building was sold before construction started.

Veidekke Eiendom is owned by one of Norway's largest construction firms. Like NCC Eiendom, it develops office buildings for sale to investors and for rental. Neither NCC Eiendom nor Veidekke Eiendom invests in special adaptability measures unless the market demands them and/or the first tenants pay for them.

5.4 MEASURES TOWARDS INCREASED ADAPTABILITY

Buildings with high adaptability will often cost appreciably more than buildings with limited adaptability. This is probably true if one aims for the highest possible degree of adaptability, including the opportunity to use a floor for any workplace design, to introduce new cells, meeting rooms or special functions wherever needed. A high degree of adaptability also means that there is little need to complement or change the technical infrastructure or call in craftsmen and specialists.

What seems to be a sensible approach is to limit the adaptability in certain parts of the building, such as fixing the areas for meeting rooms and special functions, and restricting future scenarios as to the most likely work styles and organisational changes that can be accommodated. Designing the building to take advantage of the opportunity to let or sell different parts also makes good sense.

Several expert workshops and discussions have been conducted as part of the study (Arge and Landstad, 2002). These have concluded that the list of measures in Table 5.2 should be met, if a sufficiently high level of adaptability in an office building is to be provided.

The selected building elements and adaptability measures can be challenged and, indeed, should be. They may also change over time with advances in building technology. So far in the study, these measures are viewed as the ones most suitable for discussing as to what different real estate actors do in the context of adaptability.

Individual measures have not been weighted, at least not so far. This matter can certainly be discussed, as measures that give buildings long-term adaptability could be more important than measures related to short-term adaptability. Gross floor–to–floor height is, for example, more important than 3D and zone base cabling. For the purpose of this chapter, all the measures carry equal weight.

Table 5.2 Building elements and respective adaptability measures.

Building elements	Adaptability measures
Building volume	
form	divisible for rental
width	15–17 m
Functional organisation	separation of areas with special functional demands (performance criteria) and common work areas
Floor–to–floor height	
gross	minimum 3.6 m
net	minimum 2.7 m
Technical grid/module	
dimension	2D workplace approx. 2.4 x 2.4 m
functionality	full
HVAC	
system air	variable air volume (VAV)
capacity	extra capacity
Cabling (electrical and ICT)	
service duct	raised floor
system electrical and ICT	3D and zone based
Suspended ceilings	
horizontal sound barrier	flat and soundproofed suspended ceiling
vertical sound barrier	downstand partitions above suspended ceiling
Internal walls	'plug and play'
Control systems	EIB/LonWorks
Fire precautions	sprinklers throughout

Another aspect that is, as yet, undecided, is whether or not to give priority to the measures in relation to costs and/or benefits. This is a major task in the study. The results presented here do, however, make a significant contribution to the discussion about the cost–benefits or profitability of investing in adaptability.

Different adaptability measures influence different kinds of flexibility. Table 5.3 sets out the major influences of adaptability on the three different aspects of flexibility discussed in the beginning of this chapter, namely financial, physical and functional.

Table 5.3 The influence of adaptability measures on different kinds of flexibility.

Building elements	Adaptability measures	Type of flexibility		
		Finan-cial	Phys-ical	Funct-ional
Building volume				
form	divisible for rental	•		
width	15–17m			•
Functional organisation	separation of areas with special functional demands (performance criteria) and common work areas	•		
Floor–to–floor height				
gross	minimum 3.6m			•
net	minimum 2.7m			•
Technical grid/module				
dimension	2D workplace approx. 2.4 x 2.4m			•
functionality	full			•
HVAC				
system air	variable air volume (VAV)			•
capacity	extra capacity			•
Cabling (electrical and ICT)				
service duct	raised floor			•
system electrical and ICT	3D and zone based			•
Suspended ceilings				
horizontal sound barrier	flat and soundproofed suspended ceiling		•	
vertical sound barrier	downstand partitions above suspended ceiling		•	
Internal walls	'plug and play'		•	
Control systems	EIB/LonWorks	•		
Fire precautions	sprinklers throughout	•		

The following measures are important for securing financial flexibility (corresponding to the Scandinavian concept of elasticity):

- The building should be organised in such a way that it can be subdivided into several parts that can be sold or rented independently of the others.
- Building functions should be separated so that common or special facilities can serve different tenants. This also promotes functional flexibility.
- EIB/LonWorks should be used to enable easy reconfiguration of building services when a change of tenant occurs.
- Sprinklers are required throughout (according to Norwegian regulations) in order to operate with flexible subdivision of large office areas.

The following measures are important for achieving physical flexibility (Scandinavian = flexibility):

- Flat and soundproofed suspended ceilings—a flat ceiling for the minimum number of height alternatives on internal walls and a soundproofed ceiling to secure normal sound insulation between rooms.
- Downstand partitions above suspended ceilings in predefined wall positions. Partitions are needed for rooms with high requirements concerning sound insulation.
- 'Plug and play' internal walls to enable easy and rapid alteration of wall/room configurations.

The following measures are important for achieving functional flexibility (Scandinavian = generality):

- A 15–17 m wide building provides space for cellular and team or project group workplace designs and is sufficiently space efficient.
- A gross floor-to-floor height of approximately 3.6 m is necessary to obtain a net height of 2.7 m over the whole floor, which is needed by law for permanent workplaces.
- A technical grid with a sufficient functionality for engineering services, covering the depth and length of the building, to serve permanent workplaces over the whole floor.
- A raised floor acting as a service duct for HVAC is not common in Norway. For cabling, raised flooring has been introduced, but is not in widespread use.
- Variable air volume (VAV) systems are used mostly for meeting rooms and similar areas in Norway, and not for work areas due to their cost.
- 3D and zone based cabling systems have been introduced in Norway just recently and are not in general use.

In all, there are nine measures that provide functional flexibility, three providing physical flexibility and four providing financial flexibility. The two-dimensional technical grid, with a high percentage of the units given over to full functionality, that is 100 per cent servicing, is one of the most expensive measures to implement. This is where the most difficult weighting occurs with respect to office projects—the degree of flexibility against cost.

5.5 WHAT THE ACTORS DO CONCERNING ADAPTABILITY

We have aggregated what the different real estate actors in the study do concerning adaptability and this is shown in Table 5.4. The table indicates the number of measures or the actors' degree of investment in flexibility, as well as the kind of flexibility to which different actors give priority.

Table 5.4 What different real estate actors do concerning adaptability in office buildings.

Type of flexibility	No. of flexibility measures/total, actors developing real estate		
	to serve their core business	for rental and management	for sale to investors
Financial	18/20	16/16	6/8
Functional	30/44	14/36	4/18
Physical	5/12	5/12	2/6
Total	52/76 (68%)	35/64 (55%)	12/32 (38%)

The case data show that actors developing real estate to serve their core business are investing the most in measures that make office buildings adaptable. Next are actors whose core businesses are real estate development, rental and management. Those investing the least are actors developing real estate for sale to investors.

The data also show that all real estate actors, regardless of who they are, give priority to the measures offering financial flexibility. However, the measures offering functional flexibility also influence the future market value of the building.

Functional and physical flexibility seem to be evenly ranged by the actors. If we look closer at the different actors, there are obvious differences between them. While those who develop real estate for their core business invest in two thirds of the measures offering functional flexibility, the others are investing in forty per cent or less of the measures. Investments in physical flexibility are, however, different. Here, actors developing real estate for sale to investors are investing in about one third of the measures, while the others are investing in about forty per cent.

5.6 WORKING HYPOTHESES

When the study began, there were several working hypotheses based on what could be observed in the marketplace, in the literature and according to conventional wisdom. These hypotheses were used to discuss the data from the case studies.

Working hypothesis 1

User organisations that are used to rapid and frequent changes, and that have experienced the costs and disturbances in connection with refurbishment and physical changes in the building, will be more motivated than others to invest in, or ask for, adaptability, especially functional and physical flexibility.

This is true for Telenor, but seemingly not so for Ergo Group if we look at the number of measures relating to the change rate in the core business. However, Ergo Group achieved exactly the flexibility it wanted—a universal, open workplace design that can easily accommodate more employees, as well as the opportunity to change into a conventional cellular office design. Telenor, on the other hand, knew exactly what its previous refurbishment costs were for organisational or other changes, and also the kind of flexibility it wanted. Its office buildings are without doubt the most flexible of all the cases in the study.

Among the actors who develop real estate for rental and/or sale to investors, Avantor has the most customers who are subject to change. Despite this, it is not the front-runner concerning investment in flexibility measures. ROM Eiendomsutvikling, the real estate company of the Norwegian Railways (NSB), is doing much better in that respect. One explanation for this may be that the first building that ROM was developing in Bjørvika (which is case material in this study) was supposed to be the new headquarters for NSB. The situation has changed, but the high quality ambitions have remained unchanged. Also, Bjørvika will generate a much higher return on investment and consequently provide room for higher investment costs than, for example, Nydalen, where Avantor is undertaking its real estate development.

Working hypothesis 2

Actors who develop real estate for rental and facility management will be more motivated to invest in adaptability than actors who develop real estate for sale to investors.

The study clearly supports this hypothesis, as the actors who develop real estate for rental and facility management invest in more than half of the adaptability measures whilst actors who develop real estate for sale to investors only invested in one third of the measures.

Working hypothesis 3

All actors who build new office buildings, whether for their own use or for rental or sale, will invest in measures that make the building financially flexible.

Four measures are especially important when talking about financial flexibility: building form, functional separation, EIB/LonWorks and fire sprinklers. There is no doubt these are among the measures most common in all the organisations in the study. This means that all of them, not only those who develop for rental, are conscious about changing user needs for space and the importance of market orientation and the value of the building.

Working hypothesis 4

Users who are reorganising frequently, and whose number of employees is fluctuating, will place a high priority on functional flexibility measures.

If we look at the measures that provide functional flexibility in buildings, and the users who have a very high change rate, we find no evidence that this hypothesis is true. Telenor does not have a much higher change rate than Ergo Group, for example, but it has invested far more in measures that provide it with functional flexibility. On the other side, Norsk Hydro is planning to invest in the same functional flexibility measures as Telenor, despite the fact that it does not change as frequently as Telenor.

Working hypothesis 5

Actors who develop real estate and who do not know what kind of workplace design their customers prefer will place a high priority on measures that provide the building with physical flexibility.

If we look at the measures that provide the building with physical flexibility, there is some evidence that most real estate developers prepare for several workplace solutions, i.e. open, team-based workplace designs, and traditional cellular office designs. They do so in very traditional ways, installing downstand partitions above the ceiling (for noise prevention and for connecting walls) in positions where walls may be placed, early or later on. Flat and soundproofed ceilings, combined with 'plug and play' internal partition walls, such as Telenor has chosen, demand higher gross floor-to-floor heights than commercial real estate developers seem prepared to offer.

5.7 ASSERTIONS

Only two of the working hypotheses seem to be validated in the study, namely the second and third: actors who develop real estate for rental and facility management will be more motivated to invest in adaptability than actors who develop real estate for sale to investors. All the actors in the study provided for financial flexibility. It is an easy provision, but it does cost more for vertical accesses and fire sprinklers. None of the other hypotheses are validated.

Assertion 1

All real estate actors, whether they are building to serve their own core business, or for rental or sale, are concerned about the building's market or financial value, and prepared to invest in measures that give their buildings these qualities.

If we count the number of flexibility measures, those that have the highest numbers among the cases are Telenor, both in Fornebu, Oslo and in Bergen, ROM Eiendomsutvikling (with Linstow Eiendom in Bjørvika) and Norsk Hydro.

Assertion 2

Actors who develop real estate to serve their own core business are more motivated to invest in flexibility measures than actors who develop real estate for rental or for sale.

Actors who develop real estate to serve their own core business are much more concerned about the building's use value than its exchange value in the short run. Such actors are forced by their core business to focus on the cost of use over time, and factors that can enhance the productivity of the core business. Actors, whose core business is real estate development for rental or sale, have their main focus on the building's exchange value in the short run and their own profitability.

Assertion 3

Actors who develop real estate for rental and facility management will be more motivated to invest in adaptability than actors who develop real estate for sale to investors.

Actors whose core business is real estate development for rental and facility management have experienced rapid changes in their clients' core business in the last few years, resulting in demands for new and more efficient layouts, shorter rental agreements and more frequent changes of clients than before. This has forced them to think about adaptability, both to serve their customers and to be able to make the changes without losing too much time and thus rental income.

Assertion 4

Provided that the clients, as well as their architects and other consultants, are what one might call informed, the projects would seem to influence each other.

Informed clients and consultants actively discuss and evaluate the solutions chosen by their colleagues in relation to their own projects. Given that adaptability is both a business factor and an environmental issue at the moment, there is reason to believe that 'good solutions' concerning adaptability will in time be a common characteristic in office buildings. If not, we may experience even more useless and obsolete speculative office buildings than were built in the 1980s and early 1990s.

5.8 CONCLUSIONS

The opening question was why different actors decide differently about investing in adaptability for office buildings. Our data gives us sufficient ground to say that actors who are developing real estate to support their own core business are more apt to invest in adaptability than actors who are developing real estate for sale, or for rental and management. There is also clear evidence that actors who develop real estate for rental and management are more apt to invest in adaptability than actors who develop real estate for sale.

Their different perspectives can explain this difference. Actors who build for their own core business give priority to the use value of buildings. The core business has a strong influence on what is going to be built. These actors are left with their buildings and changing user demands over time and therefore they give priority, not only to the building's initial cost, but also to the building's lifecycle cost. On the other hand, actors whose core businesses are real estate development, rental and management, will have a business related perspective on their buildings. They will give priority to the exchange value of their buildings and to their return on investment. As long as the market (representing users) does not ask for adaptability, or commercial actors do not see any benefits in providing adaptability, the latter will not offer it.

Adaptable buildings are sustainable buildings. Given that most businesses sell on their properties, one can fairly raise questions about what is going to happen with adaptability in office buildings. As Kincaid (2000) has suggested, one may ask if adaptability ought to be promoted by law, as one of several public regulations to be met in any building in line with its technical and other functional demands.

5.9 REFERENCES

Arge, K. and Landstad, K., 2002, *Generalitet, fleksibilitet og elastisitet i bygninger. Prinsipper og egenskaper som gir tilpasningsdyktige kontorbygninger* (In Norwegian), Prosjektrapport 336, (Oslo: Norges byggforskningsinstitutt).

Blakstad, S.H., 2001, *A strategic approach to adaptability in office buildings.* Dr.Ing. Thesis, (Trondheim: Norwegian University of Science and Technology, Department of Building Technology).

Gibson, V., 2000, Property Portfolio Dynamics: the flexible management of inflexible assets. In *Facility Management: Risks and Opportunities*, edited by Nutt, B. and McLennan, P., (Oxford: Blackwell Science).

Kincaid, D., 2000, Adaptation and Sustainability. In *Facility Management: Risks and Opportunities*, edited by Nutt, B. and McLennan, P., (Oxford: Blackwell Science).

CHAPTER 6

Towards a More Sustainable Industrial Workplace

John Hudson

6.1 INTRODUCTION

The *Workspace* project was a major European research initiative set up to investigate the possibility of improving the quality of production in manufacturing industry through workspace design. The workspace concept was summarised as follows:

> It is widely recognized that industrial production requires the bringing together of two systems—a technical system and a human/social system. For these two to work together effectively in a combined socio-technical system, the space in which they operate must allow them to integrate effectively. Moreover it must be flexible enough for this interface to remain, despite changes in the production process, the product and market. The role of workspace as an instrument of production is less widely recognized and hence its design and management are not subjects of strategic planning and systematic decision making, nor is the performance of workspace systematically audited (Hoefnagels *et al.*, 1998).

The project was funded under the European Commission's Brite–EuRam III programme. A network of industrial and academic partners managed by EuroFM carried out the research. Six case studies of workspace design and management from the industrial partners were undertaken. In addition to the case studies, a cross-case comparison of facilities management in each of the organisations involved was undertaken (Alexander *et al.*, 2001). The six case studies were as follows.

1. The restructuring of the test facilities of the Truck Assembly Factory at the Eindhoven site of DAF Trucks NV in the Netherlands (Hoefnagels *et al.*, 1998). The project aimed at separating the processes of inspection and repair in order to improve quality control. This involved the design and construction of a 1,600m^2 annex to an existing assembly hall. It was part of a larger project to restructure production in DAF.

2. A factory for the Rover Group at its Hams Hall site near Birmingham, UK to manufacture 4–cylinder engines (Alexander *et al.*, 1999). Unlike the other case studies this was of a completely new facility rather than the expansion or adaptation of something existing. The study was undertaken before the facility opened and focused on the design process through which workspace decisions were made.

3. The expansion of the Ericsson's Borås plant for the manufacture of electronic communication packages. The plant is situated near Göteborg, Sweden (Röjås *et al.*, 1999). The plant underwent major expansion during the 1990s in response to the growth in the market for communication systems. This expansion developed through a series of phases and resulted in the addition of around 14,900m^2 of space. At the same time that floor area was increasing, more efficient methods of space planning were introduced so that more production could be achieved in less space.

4. An office development at the ABB Transmit Oy's power transformer plant at Vaasa in Finland (Hede *et al.*, 2000). The office development was a part of a much larger project to restructure the entire working environment of the Vaasa plant in line with lean production thinking. Before the project, the configuration of the office space was such that there were significant barriers to intra-organisational communication. New single floor office accommodation was combined with new process-based teams to enhance communication and productivity.

5. A building to accommodate a ceramic multilayer capacitor (CMC) process for Royal Philips Electronics at Roermond, the Netherlands (Schless *et al.*, 2000). This is a major manufacturing plant for electronic components that has suffered competition from other plants in other parts of the world where manufacturing costs are lower.

6. The adaptive reuse of the former paint shop at the Volvo Torslanda car plant in Sweden (Kadefors *et al.*, 2000). The paint shop was a particular 'problem building' that occupied an important position within the plant. For various technical reasons it was expensive to demolish and difficult to reuse. The case study focuses on the decision making process undertaken in the adaptive reuse of the building. It reveals that a problem building can sometimes be the spur to creative thinking about the use of space.

Each of these studies produced valuable insights into the relationship between the design and management of the workplace and industrial productivity. However, the analysis of the case studies also raised a number of issues about the future economic, social and environmental directions for the industrial workplace that went beyond the remit of the initial project. In this chapter, some of these issues are explored using the general framework of sustainable development.

6.2 ISSUES OF SUSTAINABLE DEVELOPMENT

The concept of sustainable development is very broad and lends itself to multiple definitions and interpretations (Sustainable Development Commission, 2001). It is beyond the scope of this chapter to discuss them here. For the purpose of analysing the issues emerging from the case studies, a methodology known as PICABUE has been adopted (Mitchell *et al.,* 1995). This has been developed to provide indicators of sustainable development. PICABUE is not advocated as necessarily the best, or the only, method of analysis. However, it is grounded in the principles of sustainable development embodied in Agenda 21 and it has proved useful as a heuristic for understanding the issues involved.

PICABUE distinguishes four major principles of sustainable development as follows.

1. Futurity (inter-generational equity)—minimum environmental capital (resources and ecological support systems) must be maintained to ensure that the needs and aspirations of future generations are not compromised by current activities.
2. Equity (intra-generational equity)—individuals should have greater equality in access to environmental capital and should share the costs associated with human activity.
3. Public Participation—individuals should have an opportunity to participate in the decisions that affect them and the process of sustainable development.
4. Environment—the integrity of the natural environment, recognising the value of the wider ecosystem as a resource worthy of conservation, because people benefit from its use and also because it has intrinsic value beyond human resource use.

These principles of sustainable development can be used to explore the *Workspace* case studies. The following analysis is made on the basis of the published case study material. It is possible that other facts and issues would emerge from enlarged case studies that specifically examined sustainable development or environmental policies. All the companies involved, or their parent companies, have policies on sustainable development or the environment.

6.2.1 Futurity

Most of the companies in the case studies dealt with the issue of 'maintaining resources and ecological support systems' through the implementation of environmental policies and management systems as will be outlined in 2.4 below.

Futurity is about taking a long-term perspective of development. This is particularly appropriate for industrial plants that often remain in a single location for many years. The Volvo Torslanda site, for example, was established in 1958. Major industrial sites have significant effects on their local environment that can be both positive, e.g. in terms of providing employment, and negative, e.g. in terms of generating pollution. The implication of this is that decisions taken purely on economic grounds (e.g. to relocate manufacturing) could have a devastating impact upon an area.

The case study projects mainly addressed the long-term perspective in terms of the adaptability and flexibility of buildings and plant. As an example of a new build project, the Rover Group's Hams Hall plant was designed to be flexible in terms of allowing expansion without disrupting production and also to accommodate shifting production, specifically with respect to changes in product design, production volume or environmental standards. The effective reuse of redundant buildings on industrial estates can be problematic on long-lived sites as their location and dedication to particular technologies can limit their potential. The Volvo case study centred on the adaptive reuse of a redundant paint shop building and is a good example of how creative thinking can make use of a surplus 'problem building'.

The long-term sustainability of industrial plants is highly dependent upon demand for their products and relative manufacturing costs in a global market. In high consumption countries, manufacturing has declined and some companies have withdrawn from manufacturing altogether to concentrate on brand management (Klein, 2000). Although this did not occur within the time-span of the case studies it is a future possibility.

The companies involved in the case studies have not been immune to market changes and there has been considerable fluidity in ownership since the case studies were undertaken. For example, the Rover Group has been sold by BMW and as a consequence the Hams Hall site, which opened in February 2001, now produces engines for BMW. There have been major global job losses at Ericsson following a sharp drop in sales of its mobile phones. In 2001, Ericsson merged its mobile phone business with that of Sony to form Sony Ericsson Mobile Communications. In these circumstances the long-term sustainability of particular plants remains uncertain.

6.2.2 Equity

In the context of the case studies, equity can be considered both within and outside the workplace. For those in employment the workspace environment itself has a significant effect on the quality of life, an important aspect of sustainability. All of the case studies revealed a clear commitment on the part of the companies to improving the physical quality of the workplace. Ericsson, for example, provided environmentally friendly workstations and Rover developed audited workplace standards.

All the case study projects were also linked to the companies' programmes to modernise and rationalise their management approach in order to increase productivity. The impact of these programmes on the quality of work experience of the employees was less clear from the case studies than the physical quality of the workspace. The DAF study suggests that employee morale may have declined because of increases in working speed. Work by Sennett (1999) and Delbridge (1998) suggests that the introduction of new management methods may result in a decline in the quality of life for employees.

The relationship between a manufacturing plant and its locality may also have an impact upon equity. A factory may have an adverse effect upon pollution and traffic levels or on visual amenity resulting in a loss of environmental equity for neighbours. These issues are not extensively reported in the case studies.

6.2.3 Public participation

Participation can be examined both within the case study projects and in terms of the relationship between the company and the outside world. In the context of the workplace, the case study organisations were tending to move, in line with modern management thinking, away from rigid hierarchies of control towards team-based production. Within ABB, for example, hierarchical levels reduced from eight to three. In DAF, the basic element of work organisation is the team or cell of about 20 workers under a team leader. Teams often have some limited autonomy in workspace organisation and determining the division of labour.

In most of the case study projects there has been some involvement of shop floor employees in the design of facilities, although major decisions appear to have been taken at a higher level. For example, in the Ericsson project there were open workshops in which everyone, regardless of employment category, could speak their mind about the project requirements. However, it was found that decision-making 'had been a consultative, often an autocratic, process rather than a participatory process'.

The case studies did not make much mention of the participation of external stakeholders in the project process. One exception is the Volvo case study in which collaboration between Volvo and the Göteborg School Board was documented. This resulted in the use of some of the space in the old paint shop for a school project group.

6.2.4 Environment

It is in the area of the environment where most of the case study projects appear to have most clearly adopted principles of sustainable development. This is not surprising as industrial production is an obvious potential source of pollution and other environmental damage. Legislation and government guidance are perhaps most highly developed in this aspect of sustainability and there are explicit criteria and benchmarks that can be adopted.

The companies all have environmental policies and guidelines in place. However, these can be difficult to interpret in terms of workplace design and facilities management. Many have adopted environmental management systems that meet ISO 14001.

The DAF project was particularly concerned with issues of energy reclamation, sound absorption and exhaust fume filtering. Design of the Rover plant involved an assessment of environmental impact 'on land, water and surrounding environment, energy use, recycling and waste disposal, as well as emissions of air, noise and smell'. The Ericsson project considered a range of environmental issues including solvent emissions, energy conservation and recycling. The ABB study mentioned lifecycle evaluation of building materials, noise control and environmental management systems. The Volvo study mentions, in particular, energy reclamation and low chemical emissions in addition to environmental management systems. For the Philips Roermond site, there is a clear policy for attending to effects on neighbouring properties.

6.3 IMPLICATIONS FOR FACILITIES MANAGEMENT

The preceding analysis is by no means comprehensive but it does serve to demonstrate how principles of sustainable development may affect the design and management of the industrial workspace. The PICABUE framework suggests that, whilst the design and management of workspace often explicitly incorporate environmental criteria, the other three principles of sustainable development—futurity, equity and participation—are dealt with in a less structured way.

This lends support to the important role that the facilities manager can play in sustainable development. Issues such as the global shift in patterns of manufacturing may seem beyond the influence of the individual facilities manger. However, there are a number of reasons to suspect that sustainable development will become an increasingly important issue for facilities managers in the near future. These include the following.

- Changing patterns of demand as client organisations adopt policies and guidelines on sustainable development. These can include sustainability criteria for suppliers of services, which might apply to both in-house facilities managers and external suppliers of facilities management services.
- More stringent legislation—most likely in the environmental field, but also in other areas of sustainable development.
- New patterns of taxation making the polluter pay. An example of this is the climate change levy in the UK.
- An increasing awareness that principles of sustainability applied to facilities may enhance rather than detract from the 'bottom line' (Romm and Browning, 1998).

If facilities management is to develop in sustainable ways then new tools will be necessary. Sustainable development is a complex concept and the facilities manager will need to be able to evaluate the impact of individual decisions. At present there are many competing approaches to understanding and evaluating sustainable development (Sustainable Development Commission, 2001). Many potential tools can be accessed via the BEQUEST (2001) network website. Some approaches are geared to the needs of national and local government, for example the development of a broad range of sustainability indicators that measure the extent to which the country or locality in question is moving towards (or away from) greater sustainability.

Generally, these indicators are too wide to be of much use to individual facilities managers. One option would be to develop sustainability indicators specific to the sector. For example, the Dow Jones Sustainability Indexes compare companies in specific industrial sectors (DJSI, 2002). However, as Bell and Morse (1999) suggest, the development of sustainability indicators is not to be undertaken lightly and is fraught with traps for the unwary.

An alternative developed specifically for businesses is the 'triple bottom line' approach. This is based around the idea of value added, but unlike traditional accountancy approaches takes three bottom lines—society, economy and the environment (Elkington, 1997).

Whatever tools are adopted or adapted for facilities management, it is important that they are used with conviction. It is easy to be caught in the trap of adopting the language of sustainability as seen in a cynical public relations exercise known as greenwash (Greer and Bruno, 1996).

6.4 CONCLUSIONS

The *Workspace* project has shown that research can often give insights into subjects that were not a part of the original remit. In this project, the focus was on the issues of design and productivity in the industrial workplace. This has provided material for exploring the broader issue of sustainable development and facilities management. Although the analysis of the *Workspace* project findings, from the perspective of sustainability, is suggestive it is by no means conclusive. It would seem that there is a considerable potential for facilities management to contribute to sustainable development. However, a detailed programme for undertaking this task remains.

6.5 REFERENCES

Alexander, K., Weald, B., Anderson, A., Asselbergs, C. and McCaughan, R., 1999, *Production Workspace Working Paper – Case Study 2: Report on Project at Rover Group Limited, EuroFM Report, Vol. 2,* (Nieuwegein: Arko).

Alexander, K., Best, C., Östman, T. and Barakat, S., 2001, *Production Workspace Working Papers – Two Workspace Appraisals, EuroFM Report, Vol. 9,* (Nieuwegein: Arko).

Bell, S. and Morse, S., 1999, *Sustainability Indicators,* (London: Earthscan).

BEQUEST, 2001, *BEQUEST Extranet: Building Environmental Quality for Sustainability Through Time.* Available http://www.surveying.salford.ac.uk/bqextra/ (accessed 9 December 2002).

Delbridge, R., 1998, *Life on the Line in Contemporary Manufacturing,* (Oxford: Oxford University Press).

DJSI, 2002, *Dow Jones Sustainability Indexes.* Available http://www.sustainability-index.com (accessed 9 December 2002).

Elkington, J., 1997, *Cannibals with Forks,* (Oxford: Capstone).

Greer, J. and Bruno, K., 1996, *Greenwash,* (Penang: Third World Network).

Hede, J., Kohvakka, A., Leväinen, K., Veijalainen, M., Siipola, A. and Siuko, U., 2000, *Production Workspace Working Paper – Case Study 4: Report on Project at ABB Current Oy, EuroFM Report, Vol. 4,* (Nieuwegein: Arko).

Hoefnagels, W., Potters, P., van Eijnatten, F., Keizer, J. and Asselbergs, C., 1998, *Production Workspace Working Paper – Case Study 1: Report on Project Breakthrough at DAF Trucks NV Eindhoven, The Netherlands, EuroFM Report, Vol. 1,* (Nieuwegein: Arko).

Kadefors, A., Lindahl, G., Spetz, H. and Andréasson, K., 2000, *Production Workspace Working Paper – Case Study 6: Report on Project at the Torslanda Plant – Volvo Car Corporation, EuroFM Report, Vol. 6,* (Nieuwegein: Arko).

Klein, N., 2000, *No Logo,* (London: Flamingo).

Mitchell, G., May, A. and McDonald, A., 1995, PICABUE: a methodological framework for the development of indicators of sustainable development. *International Journal of Sustainable Development and World Ecology,* **2**, pp. 101–123.

Röjås, L., Törnqvist, A., Hinnerson, J. and Persson, H., 1999, *Production Workspace Working Paper – Case Study 3: Report on the Borås plant – Ericsson Microwave Systems AB, EuroFM Report, Vol. 3,* (Nieuwegein: Arko).

Romm, J. and Browning, W., 1998. *Greening the Building and the Bottom Line.* (Snowmass, CO: Rocky Mountain Institute). Available http://www.rmi.org (accessed 9 December 2002).

Schless, P., van Eijnatten, F., Keizer, J. and Asselbergs, C., 2000, *Production Workspace Working Paper – Case Study 5: Report on Building AB at Philips Components Roermond, EuroFM Report, Vol. 5,* (Nieuwegein: Arko).

Sennett, R., 1999, *The Corrosion of Character,* (New York: Norton).

Sustainable Development Commission, 2001, *Unpacking Sustainable Development.* Available http://www.sd-commission.gov.uk/commission/plenary/apr01/unpack/index.htm (accessed 9 December 2002).

CHAPTER 7

The Emergence of Supply Chain Management as a Strategic Facilities Management Tool

Margaret-Mary Nelson

7.1 INTRODUCTION

Definitions of facilities management have evolved over the years with the changing nature of the sector's work. Facilities management is believed to have first originated in the US in the 1960s with the growing practice of banks to outsource the processing of credit card transactions. In the 1980s, it emerged as 'the development, co-ordination and control of the non-core specialist services necessary for an organisation to successfully achieve its principal objectives' (US Library of Congress in 1989; see Mole and Taylor, 1992). A year later, Becker (1990) stated that it 'refers to buildings in use, to the planning, design, and management of occupied buildings and their associated building systems, equipment, and furniture to enable and (one hopes) to enhance the organisation's ability to meet its business or programmatic objectives'.

Barrett (1995) then saw it as 'an integrated approach to maintaining, improving and adapting the buildings of an organisation in order to create an environment that strongly supports the primary objectives of that organisation'. Alexander (1996) has subsequently emphasised the need for dynamism by stating that it is 'the process by which an organisation ensures that its buildings, systems and services support core operations and processes as well as contribute to achieving its strategic objectives in changing conditions'. More recently, the British Institute of Facilities Management has defined facilities management as 'the integration of multi-disciplinary activities within the built environment and the management of their impact upon people and the workplace' (BIFM, 2001). All definitions, except the first, assume that the workplace is a building in which offices are contained and make no allowance for the workspace or the mobile worker. It is assumed that the workplace is contained within a building.

Traditionally, facilities management has been seen simply as the management of buildings and building services. However, in the US Library of Congress definition reported above there is no mention of buildings. The current trend is to view facilities management as the management of non-core company assets to support and increase the efficiency of the main business of the organisation. This includes the management of buildings and building services as well as management of other support services such as porterage, cleaning, and maintenance of equipment and furniture, in order to create a working environment that supports the primary objectives of the organisation. Its goal is deemed to be organisational effectiveness, i.e. helping the organisation to allocate its resources in a way that allows it to flourish in competitive and dynamic markets (Becker, 1990).

7.2 SUPPLY CHAIN MANAGEMENT

Supply chain management (SCM) is a term commonly used in the manufacturing and retail sectors, but which is only now being examined in facilities management. Manufacturing typically views it as 'all of those activities associated with moving goods from raw-materials stage through to the end user. This includes sourcing and procurement, production scheduling, order processing, inventory management, transportation, warehousing, and customer service' (Quinn, 1997). Barrett and Sexton (1998) adopt a knowledge management view and define it as 'the explicit creation and systematic management of vital knowledge through the supply chain. It requires turning individual firm knowledge into supply chain knowledge that can be shared with key members of the network and appropriately applied to add value. The supply chain and its constituent partners will continuously adapt their behaviour to reflect new knowledge and insights.' The National Institute for Transport & Logistics in Dublin on the other hand has defined it as 'the management of all the activities in any of the companies involved in a supply chain to achieve two things: to provide the highest possible level of customer service at minimum cost' (NITL, 2000). This may have been true of facilities management in the past, but the trend in best practice facilities management is for best value to be used as main procurement criterion instead of lowest cost (CFM, 2001).

The term 'supply chain' can mean different things to different industrial sectors and individual organisations within them. It is a 'system through which organisations deliver their products and services to their customers' (Poirier and Reiter, 1996). To the facilities manager, the supply chain would be the system through which services are delivered to support the business objectives of the organisation. This covers the client, customers (as these may not necessarily be the same), users and visitors (all of whom make up the demand side of the chain), and suppliers and other collaborating parties involved in the provision of a facilities management service.

Typically, a supply chain is mapped back to the manufacturer of not just the product, but its constituent materials as well. In facilities management, this would imply the integration of the FM supply chain with the construction supply chain (Figure 7.1) to provide an integrated chain in relation to the physical (buildings) aspect. When mapping the construction supply chain, Barrett and Sexton (1998) have demonstrated that clients' requirements are fed up the chain, whilst performance is fed down the chain. Hence, decisions made at the construction stage would directly affect the building's performance throughout its lifecycle and the management of the facilities.

The term 'supply chain' can mean different things to different industrial sectors and individual organisations within them. It is a 'system through which organisations deliver their products and services to their customers' (Poirier and Reiter, 1996). To the facilities manager, the supply chain would be the system through which services are delivered to support the business objectives of the organisation. This covers the client, customers (as these may not necessarily be the same), users and visitors (all of whom make up the demand side of the chain), and suppliers and other collaborating parties involved in the provision of a facilities management service.

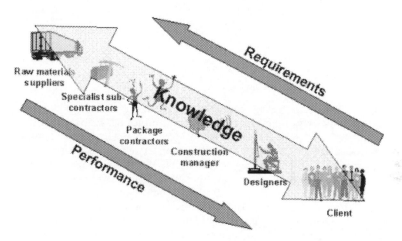

Figure 7.1 Construction supply chain (Barrett and Sexton, 1998).

The potential for change is greater at the conceptual stage of a building project, because the least cost of change occurs here—Figure 7.2. However, it is normal for changes to be carried out after the handover of the completed building, where the least potential exists and where the highest costs can be incurred.

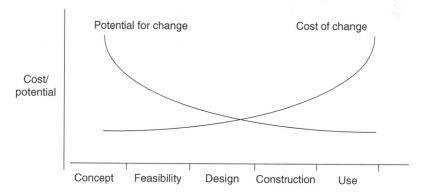

Figure 7.2 Typical building project lifecycle (Source: Gallicon presentation).

Brand (1996) found that all buildings are forced to adapt. He quotes from Sullivan in 1896 where 'form ever follows function', which would suggest that although the cost of change when the building is in use is higher than at the concept stage, it is natural that the building should continually adapt to meet the current needs of occupiers.

The strategic role of facilities management can be seen as containing sufficient knowledge to influence the performance of the building if activated in the early design stages. The facilities manager should have knowledge of the organisation's business and its physical and spatial needs, as well as physical resources available to meet those needs. In this way, facilities management can play a strategic role in influencing business decisions related to the organisation's physical requirements. How then does supply chain management add value to this process?

7.3 SUPPLY CHAIN MANAGEMENT AND FM

Alexander (1996) stated that facilities are 'an organisation's second largest expense and can account for as much as 15 per cent of turnover'. He went further to say 'they are also the largest item on the balance sheet, typically over 25% of all fixed assets'. Research has shown, however, that most organisations neither manage their facilities assets effectively nor consider facilities management as part of their overall business strategy. It is generally viewed by business organisations as a non-income generating overhead. In a 1999 BIFM survey of its members, just over 20 per cent believed that their board of directors (or similar) perceived facilities management to be of high strategic importance. Although 58 per cent believed that facilities management was perceived as having high or medium strategic importance, the majority (over 70 per cent) voted for medium to modest strategic or operational importance.

The general view is that facilities management has a supporting role to the core business of the organisation; hence, it has been difficult to argue that it should play a more strategic role in achieving the organisation's business objectives. However, in order to implement and derive the optimum benefit from SCM initiatives, facilities management needs to be viewed and addressed at the strategic level.

In a context analogous to supply chain management, facilities management can be seen as a function or series of linked activities involving the co-ordination of all efforts relating to the planning, design and management of an organisation's physical resources (Becker, 1990). In this respect 'physical' includes spatial, environmental, human and financial resources. Nutt (1992) has stated that:

- its focus is on post-occupancy rather than pre-occupancy issues
- the central rationale is management decision and implementation
- responsibilities cover all of the five primary types of resource—physical, spatial, environmental, human and financial
- the concern is with an integrated approach and it does not concentrate on any particular part of the problem field.

The current view is that facilities management should not only focus on post-occupancy issues, but should also be actively involved in pre-occupancy issues. The reality of the facilities management market has, however, been that it has not taken an integrated approach to supporting the business objectives of the organisation, although it is increasingly moving in this direction. Barrett (1995) saw the key aspects of facilities management as an integrating role in which management issues predominate over technical matters, and a service justified and oriented towards making a positive contribution to the primary business. Its main focus, however, has been service delivery. New initiatives in the construction sector such as the generic design and construction process protocol (Wu *et al.*, 2000) emphasise the need for integration and the role of facilities management in the construction supply chain.

An examination of facilities management functions is required in order to appreciate the potential for supply chain management in facilities management. Then and Akhlagi (1992) classified facilities management into three categories: strategic, tactical and operational. They argued that responsibilities, roles and tasks at each of these levels are different, and that facilities management decisions should be taken at board level and integrated into the overall business plan. These decisions, which would affect the corporate structure of the organisation, are then translated into tactical issues and, finally, into operational activities. For example, if a strategic decision to outsource the provision of facilities management services has been undertaken by the organisation, it would need to be reflected in the organisational structure and the management of FM and supply chain issues by the organisation. In-house roles and tasks would be different in an organisation that has outsourced the provision of its facilities management services, and this needs to be reflected in the organisational structure.

Supply chain management is also concerned with the elimination of waste (or non-value adding activities) from the total acquisition costs of, in this case, facilities management. The business case for effective management of an efficient supply chain is backed-up by research, which shows that '60% of variable costs are driven by decisions made outside the organisation' (Ross, 1999). This highlights the importance of supply chain management. Effective management of the supply chain would enable it to be optimised, as well as allowing control over 60 per cent of the organisation's variable costs. Considering that facilities account for as much as 15 per cent of an organisation's turnover, this would imply that at least a 9 per cent proportion of the organisation's turnover can be effectively controlled through managing the supply chain. Hammer (1998) underscores this point by stating that supply chain reengineering was the next logical step from business process reengineering (BPR) and that if you 'improve the total system, everybody comes out ahead'.

7.4 DRIVERS FOR CHANGE

The growing interest in supply chain management issues in facilities management has come about from a combination of market and client/customer forces, namely changes in the market and competition, regulations and legislation, and technology.

These drivers are similar to those of other industrial sectors that have adopted supply chain management (Braithwaite, 1999).

Five factors have been identified as stimulating the growth of facilities management (Becker, 1990):

1. employee expectations
2. cost of mistakes
3. global competition
4. high cost of space
5. information technology.

Changing regulations and legislation should also be added to this list based on experiences from other industrial sectors (Braithwaite, 1999). In order to manage these stimuli effectively, facilities management processes need to be integrated and well managed. They also need to be integrated with the overall strategy of the business.

These stimuli have resulted in the requirement to approach facilities management from other than the direction of traditional methods and, in this way, are driving the move towards supply chain management. Employee expectations should be regarded as stakeholder expectations in order to widen the view, and information technology should be seen as an enabler rather than a driver. Last, the facilities management market is evolving continually. It has moved a long way from mere delivery of 'hard' services, to a business support service.

Research is required into the real cost of mistakes to an organisation. This includes not just the direct and indirect financial aspects, and the effect of mistakes on workplace productivity, but also human and social aspects. For example, how does the workplace environment affect employee motivation, and how do employees respond to the measures implemented to correct mistakes?

The sector has also witnessed a gradual move away from lowest cost to best value. Both public and commercial organisations now make procurement decisions based on the concept of value for money. As competition erodes profit margins, supply chain costs become a critically important component of the bottom line (Beyer and Ward, 2000).

Increasing consumer power has resulted in customer satisfaction becoming a major driver for change. The adage 'the customer is king' is even more powerful in a service sector where the stakeholders, who constitute the customer, are many and varied, and demand the satisfaction of as many and varied 'wants'. The ever-changing demand for the provision of more sophisticated facilities management services has meant that facilities management needs to be more dynamic, or risk losing out to other professionals who are willing and able to provide these services.

New procurement initiatives including just-in-time, outsourcing and partnering are challenging traditional methods. Just-in-time, for example, has brought about a change from stockpiling and staff retention, to procuring (services, materials and human resources) as and when needed. Dell Computers is a good example of how successful this procurement method is when well managed, and 'demonstrates the value and effectiveness of supply-chain integration' (PRTM, 1999). On the client side of facilities management, some people view just-in-time as the way forward in procuring facilities. Herein lies a challenge to facilities management providers to deliver facilities on such a basis.

Outsourcing has opened up the facilities management sector to competition, and encouraged benchmarking and best practice. Competitive advantage on an organisational level is much more than doing things better than your competitors, but staying ahead of them. Facilities managers are being challenged by other professionals, and have seen business development managers and web-enabled virtual supply chain networks encroach upon their roles and responsibilities. The former demands an increasingly multi-skilled professional. The latter is a direct consequence of advancement in technology, which has not only resulted in changes to the nature of the facilities being managed, but also to the way they are managed.

The financial markets have traditionally held a poor view of construction companies. The need to increase investors' confidence and enhance shareholder value has led to a number of listed construction firms moving into facilities management and remodelling their image at the same time. Return on investment is a prime motivator for change in these companies. Supply chain management can be seen as an image enhancer for organisations that want to be seen as modern and innovative, and delivering to time and within budget.

Changing regulations and legislation, especially as a result of European law, and the drive towards environmental friendliness, promoting efficiency in fuel use, recycling and certification, and noise and congestion reduction have also had their impact on facilities management. Many client organisations are now highlighting environmental issues and have firmly placed them on their boards' agenda.

UK government directives on procurement such as outsourcing, partnering, prime contracting and the Private Finance Initiative (PFI) are also a major driver for change. The government, as part of its Achieving Excellence Programme for public procurement, has directed that 'From 1st June 2000 all Central Government clients should [...] limit their procurement strategies for the delivery of new building to PFI, Design and Build and Prime Contracting' (Holti *et al.*, 2000). Its belief is that all three procurement strategies 'can only achieve best value for money if they are based on the integration and management of the supply chain' (Holti *et al.*, 2000).

7.5 SUPPLY CHAIN MANAGEMENT ISSUES IN FM

In a study of supply chain management issues in facilities management (Nelson, 2001), no consensus was found amongst practitioners as to the most important supply chain management issues in facilities management. As the concept is emerging in facilities management, there is little literature or documented evidence available in this field. Organisations consider it as a strategy for competitive advantage that they are not too willing to share with others.

The study's literature review revealed that important issues are flexible working (including changes in working practice), outsourcing, benchmarking, cost-effectiveness (including value for money and efficiency), space planning, the environment, energy management and HR matters (BIFM, 1997). These are, on the whole, very similar to the findings from the study. Issues were also different depending on whether the organisation was a client or provider. Amongst the issues raised were the following.

1. Capturing client's requirements and the specification of service delivery standards.
2. Linking facilities management with the strategic business processes.
3. Technology and its effect on working patterns—technology has not only led to a change in the way we work, but where we work and the way buildings are constructed to accommodate technology. The growing influence of the Internet and web-enabled technology is further impacting on facilities management. Technology should not, however, be seen solely as a driver, but as a means for supporting business needs.
4. Process capture is also an issue as many organisations are not mature enough to capture their processes adequately then translate them into an e-format, including all necessary safety checks.
5. Consolidation or rationalisation of the supply chain—client organisations have this at the top of their agenda. Most deal with the issue by the appointment of key suppliers or a managing agent. It has, however, been opined that, rather than reducing the number of suppliers, it is more important to have a clear brief, clarify grey areas and ensure consistency of supply. Organisations with scattered regional offices are moving towards reducing the number of local suppliers and appointing a national supplier for all their offices. This leads to a loss of local feel and knowledge and does not necessarily improve service or management. The question should therefore be—who are our suppliers and how can we improve relationships?
6. Managing environmental issues in service delivery, including embedding these in service delivery specifications.
7. Alignment of organisations' business processes—organisations need to address internal alignment issues and have an infrastructure to support their supply chain management and, in particular, relationships with service providers. Supply chain initiatives fail where the client organisation does not have an internal supporting infrastructure for its procurement initiatives.

Further studies have shown that the specification of standard codes and protocols are also an issue in the facilities management market. These issues range between technological and benchmarking protocols. Technologies employed by organisations will vary within the supply chain and may be incompatible. In recent benchmarking studies (CFM, 2001), the lack of a common benchmarking protocol employed in facilities management was highlighted. This leads to two sets of problems. The first are issues related to compatibility of information technology systems, and the second are issues to do with comparing best practice.

The facilities management market is also continually evolving, so that 'yesterday's competitive advantage is today's competitive necessity' (Hammer, 1998). The sector now refers to infrastructure management and asset provision, whereas asset management was once the current term. E-business and e-procurement are also changing many facets of facilities management. Bacon (2000) believes that e-business provides an opportunity for facilities management to streamline complex processes and the exchange of information. Areas of opportunity in e-business for facilities management include the exchange of information, procurement, resource management and business development.

7.6 BUSINESS BENEFITS

Earlier it was stated that facilities are 'an organisation's second largest expense and can account for as much as 15% of turnover' and 'they are also the largest item on the balance sheet, typically over 25% of all fixed assets' (Alexander, 1996). However, facilities as an item are mostly perceived as an overhead or non-income generating expenditure. This aspect of business is increasingly being outsourced in a bid to reduce costs and focus on core activities.

Supply chain management involves long term strategic planning and is a method of managing the delivery of services effectively and efficiently to deliver 'faster, better and cheaper'. Experiences from other industrial sectors show the benefits of implementing supply chain management strategies, which include the following.

- Better knowledge and understanding of market trends and key customers needs
- Better understanding of suppliers capabilities
- Enhanced problem solving
- Faster response times
- Flexibility and a strategic capability to plan and innovate
- Improvement in forecast accuracy
- Improvement in inventory reduction
- Lower operating costs and reduction in real costs
- More efficient and effective use of technology
- Shared risk
- The development of an environmentally sound service
- Waste reduction and more effective use of resources and skills
- Free up of working capital realised through supply chain management efficiencies
- Maximisation of asset return
- Increased capacity to innovate within the supply chain as knowledge is shared along the supply chain
- Opportunity to refocus on the interface with users.

These benefits address the facilities management business drivers and issues in the sector as identified earlier. They will lead to improvements in quality and delivery performance and increased repeat business with key customers. They can also be translated into greater confidence for long term planning, provision of greater certainty of costs, and improvements in value for money.

AMEC Facilities has, for example, found real benefits in undertaking supply chain management, including the following (O'Halloran, 2001).

- 15–20 per cent cost reduction
- 96 per cent customer satisfaction
- Reduction in lead times and fault levels
- Improved service at reduced cost
- Financial control
- Service linked to business drivers
- Payment linked to performance Full commitment to a service culture

- A forum for intellectual exchange
- A common vision and goal
- A virtual company concept
- Mentoring at company and individual level
- Rewarding success
- Embracing change as a way of life.

AMEC Facilities has been able to achieve cost reductions in the range of 16–40 per cent for its clients, whilst improving the service offered (O'Halloran, 2001). Examples such as this suggest that supply chain management does add value, not just to the client organisation but also to the service provider. It is an obvious 'win–win' situation.

7.7 CONCLUSIONS

Supply chain management, as a concept, is highly relevant to facilities management, as the latter involves the management of a network of suppliers or several supply chains. It is being driven by various stimuli, which are also driving the evolution of facilities management as an industrial sector. These include stakeholder expectations, financial expectations, market and competition, and legislation and regulations.

Supply chain management is a complex and far-reaching issue in facilities management. It impinges on all facets of facilities management and has the potential to add value to its functions. For this reason, it should be seen in its totality, with its top-level requirements and standards established before tactical and operational issues are addressed. It should also be regarded as a strategic tool for achieving competitive advantage.

The main supply chain management issues in facilities management were found to be the specification of service delivery standards, consolidation and rationalisation of the supply chain, use of technology in service delivery, environmental matters and alignment of the internal value chain and the supply chain. Overall, these issues are very similar to the results of a BIFM survey carried out in 1997.

Integration is also seen as an issue in this area. It has been suggested that the capacity for innovation in a supply chain increases with the level of integration of effort and strategic alignment. Technology is viewed as both a driver and an enabler. Changing work patterns and building types are driving the changes, whilst technology enables new ways of delivering facilities management services and building management systems. Technology is also seen as a means for achieving integration if it can be managed properly.

The benefits of supply chain management in facilities management have been emphasised by both public and private sector organisations. Benefits derived in other industrial sectors, such as manufacturing and retail, are very similar to those in the facilities management sector. A better understanding of customers' needs has been shown to lead to increased repeat business. Lowering costs whilst improving quality leads to waste reduction and the maximisation of asset returns, whether physical, technological or human assets. Risk is also shared and managed through the supply chain.

Although the organisations involved in this study agreed about the importance of supply chain management, they did not agree on what the main issues were. Most evidence was anecdotal and very few hard facts were provided to support benefits derived so far from supply chain management strategies. Gaps remain in both research and development in this area, as well as in developing a generic supply chain management tool or model for use in the facilities management sector.

7.8 REFERENCES

Alexander, K., 1996, Facilities Management. In *Facilities Management: Theory and Practice,* edited by Alexander, K., (London: E & FN Spon), pp. 1–13.

Bacon, M., 2000, Presentation on facilities management, University of Salford.

Barrett, P., 1995, *Facilities Management: Towards Best Practice*, (Oxford: Blackwell Science).

Barrett, P. and Sexton, M., 1998, *Integrating to Innovate: Report for the Construction Industry Council,* (London: Construction Industry Council).

Becker, F., 1990, *The Total Workplace: Facilities Management and the Elastic Organisation*, (New York: Van Nostrand Reinhold).

Beyer, D. and Ward, J., 2000, *Network Server Supply Chain at HP: A Case Study*. Available http://www-uk.hpl.hp.com/techreports/2000/HPL-2000-84.html (accessed 9 December 2002).

BIFM, 1997, *BIFM Survey 1997,* (Saffron Walden, Essex: British Institute of Facilities Management).

BIFM, 1999, Survey of Facilities Manager's Responsibilities, BIFM Members Survey, September, (Saffron Walden, Essex: British Institute of Facilities Management).

BIFM, 2001, *Facilities Introduction.* Available http://www.bifm.org.uk/index.mhtml?get=face/facilitiesintro.html (accessed 9 December 2002).

Braithwaite, A., 1999, Supply Chains in the 21st Century, Presentation at the 1999 Supply Chain World Conference and Exposition, Amsterdam. Available http://www.supply-chain.org/downloads/Albright.ppt (accessed 9 December 2002).

Brand, S., 1996, *How Buildings Learn: What happens after they are built*, (New York: Viking Penguin).

CFM, 2001, Financial Forum Benchmarking Project Report 2001, Unpublished report for CFM Foundation Members, (Salford: Centre for Facilities Management, University of Salford).

Hammer, M., 1998, Reengineering the Supply Chain: An Interview with Michael Hammer, *Supply Chain Management Review.* Available http://www.manufacturing.net/SCM/index.asp?layout=articleWebzine&articleid=CA155060 (accessed on 9 December 2002).

Holti, R., Nicolini, D. and Smalley, M., 2000, *The Handbook of Supply Chain Management,* (London: CIRIA and The Tavistock Institute).

Mole, T. and Taylor, F., 1992, Facility Management: Evolution or Revolution. In *Facilities Management: Research Directions*, edited by Barrett, P., (London: Surveyors Holdings).

Nelson, M., 2001, Supply Chain Management in FM, Report for SPICE FM submitted to the EPSRC, (Salford: University of Salford).

NITL, 2000, *European Driving Licence in Supply Chain Management*. Available http://www.nitl.ie/ (accessed 20 June 2000).

Nutt, B., 1992, Facility Management: The Basis for Applications Research, In *Facilities Management: Research Directions*, edited by Barrett, P., (London: Surveyors Holdings).

O'Halloran, M., 2001, Presentation on supply chain management. (London: AMEC).

Perks, H., 2000, Marketing information Exchange Mechanisms in Collaborative New Product Development: The Influence of Resource Balance and Competitiveness. *Industrial Marketing Management*, **29**(2), pp. 179–189.

Poirier, C.C. and Reiter, S.E., 1996, *Supply Chain Optimization: Building the Strongest Total Business Network*, (San Francisco: Berrett-Koehler).

PRTM, 1999, *Managing your supply chain in the 21st century*. Available http://www.e-global.es/011_prtm_managing.pdf (accessed 9 December 2002).

Quinn, F.J., 1997, What's the buzz? *Logistics Management*, (February), pp. 43–47.

Ross, K., 1999, *Creating Value from Business to Business Integration*. Available http://www.ascet.com/ascet/wp/wpRoss.html (accessed 20 June 2000).

Then, D.S.S. and Akhlagi, F., 1992, A Framework for Defining Facilities Management Education. In *Facilities Management: Research Directions*, edited by Barrett, P., (London: Surveyors Holdings).

Wu, S., Lee, A., Aouad, G. and Cooper, R., 2000, Process Protocol Toolkit, an IT solution for process protocol. In *Proceedings of ECPPM'2000, Lisbon, September 25–27, 2000*, pp. 158–193.

PART THREE

Performance: An Introduction

Jan Bröchner

Read as a whole, the two contributions in Part Three give an overview of the major issues encountered when performance is to be measured and improved in the context of facilities management. First, John Hinks in Chapter 8 explains the particular challenges that present themselves to anybody who has to manage the performance of a support service, as opposed to the management of a core business. Attempting to measure the performance of facilities and of facilities management while disregarding the effects on the core activities of a firm leads to a gap between business managers and facilities managers. Bridging this gap implies the development of more sophisticated measures that are able to trace how support performance leads to core business performance. In Chapter 9, Dilanthi Amaratunga and David Baldry illustrate how a set of non-financial indicators of performance for facilities management can be devised. Their example is taken from higher education facilities, but their way of adapting the Balanced Scorecard approach tackles many problems that emerge when performance measurement is introduced in a facilities context.

While the 1990s can be thought of as the decade of performance measurement, it is mostly so because the decade saw the collapse of an excessive reliance on financial measures derived from accounting data (Eccles, 1991). In a sense, we have returned to the skills of premonetary societies such as found in ancient Mesopotamia (Nissen *et al.,* 1993). Performance measurement and management is thus age-old, and the first clearly identifiable discussion of measurement practice for facilities management is related to the outsourcing of aqueduct maintenance in Rome; Frontinus as commissioner for the system relates with pride in his handbook that 'Both of these large gangs, which regularly were diverted by exercise of favouritism, or by negligence of their foremen, to employment on private work, I resolved to bring back to some discipline and to the service of the State, by writing down the day before what each gang was going to do, and by putting in the records what it had done each day.'

Like duty, loyalty, value and management itself, performance is a word that carries a connotation of the world of chivalry, having passed from medieval French into the English language. In common with 'facilities', these words often resist translation into languages spoken in nations with a less pervasive history of feudal relations. Also, these concepts sometimes sit uneasily with concepts derived from mainstream economic theory. The recent wave of performance measurement should partly be explained by a fresh injection of feudal thinking, misunderstood or not, with Toyota supply chain management as the primary paradigm. The global pre-eminence of Japanese manufacturing during the 1980s led to considerable interest in establishing closer relationships between firms, and also to an understanding that replacing market prices by partnering implied a need for broad measures of performance, if efficient resource use was to be monitored. In addition, and perhaps equally important, advances in information technology, especially the emergence of large scale use of networked personal computers, created the practical possibilities for more refined measurement of performance.

While these impulses came mostly from the manufacturing sector of the economy, there was also an inheritance from the 1960s in the shape of building performance studies. Information technology and mobile telecommunications here played another role by breaking up our conventional ways of looking at what constitutes efficient workplace design. As can be seen elsewhere in this volume, we are only beginning to grasp the significance of these changes in the workplace and the consequences for measuring performance.

REFERENCES

Eccles, R.G., 1991, The performance measurement manifesto. *Harvard Business Review,* January–February, pp. 131–137.
Frontinus, 1997, *The Stratagems; The Aqueducts of Rome,* tr. Bennett, C.E., repr. of 1925 ed., (Cambridge, MA: Harvard University Press).
Nissen, H.J., Damerow, P. and Englund, R.K., 1993, *Archaic Bookkeeping: Early Writing and Techniques of Economic Administration in the Ancient Near East,* (Chicago: University of Chicago Press).

CHAPTER 8

Business-Related Performance Measures for Facilities Management

John Hinks

8.1 INTRODUCTION

Business managers will continue to overlook the strategic potential of facilities management as a component in the value chain until the terms that facilities managers use for discussing its performance relate more directly to the core business drivers. Meanwhile, the normal approach to facilities management benchmarking—that of focusing on generic, data-driven, quantitative and predominantly facility-oriented measures—places the emphasis on the common denominators of average performance issues. It does not allow the value of facilities management for particular businesses to be differentiated, and it does little to help Facilities Managers relate their performance to the preoccupations of their core business clients. Furthermore, if the hackneyed saying 'what gets measured gets managed' holds, contemporary facilities management performance management may be at best a diversion, and at worst be an obstacle to the realisation of facilities management as a strategic business asset. The use and abuse of Key Performance Indicators (KPIs) and performance measurement in general have thus to be analysed.

8.1.1 Value for the business

Gauging the business value of facilities and facilities management remains a difficult issue for the world of business. The challenge of looking beyond facilities or business support services generally as cost centres, and in seeing them and their management as value-adding for the overall business appears to remain alien to the business mindset, and the management of facilities continues to be overlooked in many business school curricula.

The world of facilities management seems alert to the need to demonstrate value over-and-above cost if this business mindset is to be broadened, and committed to doing this through integrating performance measurement with the design and review of their business support services. However—and this is the main point to be made in this chapter—the combination of traditional facilities management benchmarking using common, facility-oriented indicators, plus the primacy that the comfort of numbers lends to quantifiable data, added to a tendency to measure just the inputs that facilities managers control, has led to a concentration on measurable facility-oriented issues, rather than measures that represent the performance endpoint, the impact of facilities management on the key business (Hinks, 2000a). Nevertheless, there remains a lack of agreement across the field of facilities management as to the nature of, and terminology for, performance indicators, which is hardly surprising given the recognized impracticality of creating a single comprehensive set of indicators of performance.

8.1.2 Two worlds

The distinction I am making is that while the world of facilities management assembles the case for its worth using evidence of how cost-effectively it achieves what it does, the business view of the value of facilities management will be derived from demonstrations of what effect facilities management actually has on the business. At best, this is a distinction between outputs in the form of facilities management and outcomes in the form of business results, but too often performance measures for facilities management focuses narrowly on inputs to the delivery of facilities-related services, and such measures are even more remote from the endpoint of the business value chain. As usual, the measurement of outcomes can be difficult, especially where subjectivity matters. Smith (1995) has noted the problems with achieving consensus on what the objectives and output of an organisation may be, especially with public sector or not-for-profit organisations, and also in consistently interpreting any measures that are set up to gauge this.

Bordass *et al.* (1999) have observed that relatively minor facilities management issues which may not be high on anybody's agenda have been found to have major effects on achieved performance, but current approaches to facilities management performance analysis using generalised benchmarking criteria and mass datasets, may indicate only whether a particular function is above or below an industry or sector average. Without further interpretation, these approaches do not allow us to see how a function is balanced to the particular business and its facilities management needs (Amaratunga *et al.*, 2000).

8.2 AN EXPERIMENT

At the Facility Management Association Ideaction 2001 Conference in Melbourne, participants were asked to rank the top 3 FM Key Performance Indicators (KPIs) according to effort put into data collection, and the top 3 business KPIs for their direct customers. Responses where weighted according to rank (top priority = 3x, second = 2x) and can be studied in Table 8.1, which shows clear differences in KPI priorities for a facilities management view and a business view. Note the emphasis on cost effectiveness and functionality among the FM KPIs, compared with business growth and revenue/profit in the set of prioritised business KPIs. Customer satisfaction is ranked highly both as a FM KPI and a Business KPI, whereas many other types of performance indicators show clear differences in priority.

This experiment shows that many practitioners are aware of the need to distinguish between facilities management and business performance indicators. The challenge is to explore how we can link indicators that belong to the facilities management perspective to those that are relevant to the business itself.

Table 8.1 Facilities management and business priorities for indicators.

Performance indicator	Facilities management priority	Business priority
Cost effectiveness	93	29
Customer satisfaction	82	110
Functionality	66	12
Revenue/profit	39	97
Service quality	36	3
Staff satisfaction	27	22
Space management	26	11
Business growth	21	51
Safety	17	6
Flexibility	17	6
Total effectiveness	11	6
Service response time	11	8
Synergy with company's strategic plan	11	21
Efficiency	9	3
Risk management/Technology	8	4
Change	6	3
Efficient workforce	4	6
Benchmarking	4	1
Research	3	5
Quality environment	3	20
Communication	1	2
No harm to environment	1	6

8.3 THE NARROW FACILITIES MANAGEMENT APPROACH

Historically, facilities managers have tended to measure performance from an operational efficiency perspective – referring to data which are facilities-oriented and focusing on the efficiency of the facilities spend – factors such as running costs, energy consumption, maintenance costs and rental costs per square foot. This works well for prioritising within the operational management of the facilities, and for checking how the outlays relate to the norms for the industry, budgetary targets or legal requirements. The data also tend to reduce the assessment process to checking for deviation from the lowest common denominators of running facilities efficiently.

What the data do not do is illuminate the potential competitive edge of tuning the facility to the business process. They focus attention on averages, not excellence. Nor do the data support any analysis of the correspondence between the facilities management service as a lever and the strategic priorities facing managers of the core business. So, whilst these conventional measurements allow the facility manager to assess outlay against budget, and even to compare this with the industry norm, they do not make it clear whether the organisation is spending the right amount for its needs, or whether it is getting maximum support advantage or potential flexibility. Hence it is difficult to express effectiveness and deep value, or to inform strategic business planning decisions with credible statements about effectiveness, expressed in cost control terms.

The most frequently used metrics, as found by Massheder and Finch (1998), are related to occupancy cost and operational space. All these are readily available and quantifiable. However, Kincaid (1994) has warned that facility-related benchmarking indicators can place excessive emphasis on costs. Massheder and Finch (1998) suggest that specific standards of measurement, or metrics, are essential for ensuring a common understanding of performance and to identify performance gaps. But they also conclude that facilities management metrics are not being used to reflect the overall performance of the business, and that as a consequence of this, the significance of facilities decisions on overall business success will continue to go unnoticed until a more holistic approach is applied to benchmarking.

The creation of large databases, like those of the IFMA in the USA, leads to a communal source of quantitative performance data, often in the form of indices and trends, and have arguably contributed to the quantitative emphasis for benchmarking. The negative consequence of this emphasis is a beauty contest mentality, and the temptation to engage in the quest to claim number one ranking, and perhaps more significantly, to avoid the embarrassment of an unfavourable rank, threatens (Ammons, 1999). Simply checking compliance with large datasets will only indicate whether a function is keeping up with the herd. Indeed, the ease of acquiring and interpreting information appears to be one of the driving characteristics of facilities management benchmarking (McDougall and Hinks, 2000a; 2000b).

8.4 BROADENING THE SCOPE FOR FACILITIES MANAGEMENT INDICATORS

This focus on narrow facility management issues directly conflicts with the push to widen the scope of facilities management by addressing issues like customer satisfaction, and service standards (McDougall and Hinks, 2000b). This is where performance measurement benefits from earlier investigations of building performance as perceived by building users (Leaman, 1995, Leifer, 1998, and Preiser, 1995). For customer satisfaction, it is vital to recognise the importance of particular issues to your customer, and then to consider how well your service is supporting these needs (van de Vliet, 1997).

8.4.1 Workplace effects on employee productivity

From the viewpoint of the business, the links between the facilities for work and employee productivity are important but far from easy to structure and measure. These effects have been recognized by a number of investigators. Already Oldham and Brass (1979) pointed out that specific procedures would be needed to correct the adverse effects that open offices might have on job characteristics and employee work outcomes.

In his review of earlier studies into the reactions of employees to their open plan work environments, Hedges (1982) outlined the complexity of non-facilities issues influencing satisfaction with workspace: 'numerous problems affecting employee's work were found, such as lack of privacy and high levels of distractions and disturbances [...] for a majority of employees performing managerial and technical tasks, there appears to be an inverse relationship between their satisfaction with work and their satisfaction with office conditions. Those staff with the most complex, demanding, and satisfying jobs also tended to be those most sensitive to their work environment, and also those who expressed the highest levels of dissatisfaction with office conditions.' He went on to note that these and other problems 'appear to be instrumental in creating the perceived inequity between desired and actual work output which many employees reported'. Davis and Szigeti (1982) pointed to a dearth of comparative empirical data about the physical work environments, something which is still true.

A decade later, Carlopio and Gardner (1992) noted that it was 'important that researchers examining the influence of physical environment on employees broaden their views of the physical work environment'. They went on to say that theory and speculation in these areas seemed to be far in front of our empirical knowledge base. However, they also observed, like Hedges (1982) did before, that perceptions of the work environment might be more strongly related to satisfaction for managers, professionals, and supervisors than for clerical employees or non-supervisors. So whilst like others their work illuminates a complexity of non-facilities, non-FM issues at play in overall workplace satisfaction, and only part allied to comfort per se, their studies also could not correlate workplace satisfaction with business performance.

And who are we to measure the 'sheer anonymity' of call centre buildings and the business effects of an architecture that 'tells the workers they are nothing special' (Baldry, 1997)?

Ideally, we would like to be able to answer questions such as 'how much more comfort would be derived from allocating more space per person, per task?' or by spending more on various aspects of the flexibility or service provisions of the workspace? And precisely how would the change in comfort from this reduced space detract from the performance of the business? Note also that the first question interests the facilities management directly but business indirectly, whereas the second question interests business directly but facilities management indirectly. And, as Leaman *et al.* (1999) note, the influence of the context in the performance of the workplace can be convoluted: 'hundreds of variables are involved, many of them correlated with each other, so it is usually difficult to know which to choose—some are concerned directly with the building, its people and occupying organisation/s, others with the social and technical background which is always subtly volatile, so it can be a challenge to know which variables will be significant in any situation.' And they point to the difficulty in appraising workplace success: 'business are often unclear about what they are trying to achieve, taking refuge behind "flexibility" without understanding what they are asking for, so often they get things that they didn't know they want!'

8.4.2 Grey data and facilities management performance indicators

Most business decisions are based on a combination of hard data and soft information. Intuitive synthesis has been found to be positively associated with organisational performance in an unstable environment, but negatively so in a stable environment (Khatri and Ng, 2000). Falconer and Hoel (1997) have used the concept of 'grey data' when analysing how managers deal with occupational risks.

Falconer (2002) notes the difficulty with controlling risks in work processes, noting that occupational injuries and illnesses are frequently due to failures in work systems. These systems often fail to some extent rather than completely, and signals of failure are often hidden, as many work systems are complex, dynamic, opaque, and uncertain. How are these issues reflected in current facilities management performance assessment? Five examples of the need for recognising grey data as a base for facilities management decisions should be looked at more closely: flexibility, safety, space utilisation, maintenance management and finally the greyest of all, value for money.

Flexibility

At the higher level of facility indicators, the degree of physical flexibility has been seen as a major characteristic of facilities. In particular, the advances in information technology and telecommunications that allow hotdesking, hotelling and virtual working has meant that in business sectors based on office work processes, flexibility to accommodate new patterns of working, as well as rescaling of departmental sizes to accommodate employment growth or shrinkage are critical operational aspects of flexibility in the facility. This concern with flexibility is also related to the delivery of facilities management services. However, there is a bigger, more direct business issue: does flexible working work? The answers to this question will not be satisfying if they are founded only on cost indicator data. Work flexibility is a more complex issue.

Provision of a safe environment

Measuring the provision of a safe environment would seem to be straightforward, but is actually a good example to raise the difficulties of criteria and the choice of thresholds for acceptable performance. Are the statutory minima going to be the target, and frequency of incidents going to be the metric? Or perhaps seriousness of incidents, or perhaps the repetition of any incidents? The organisation may have its own safety standards over and above the statutory minima, and because of some aspect of the process, there may be particular emphasis on some aspects of safety. This could also relate to emissions and the safety of third parties as well as operatives. Perhaps safety may be interpreted as the provision of regulatory equipment. How would you relate training in safety to the assessment of such an indicator in your organisation? Bear in mind that the term safety is a relative concept anyway, and the outlook for a simple set of performance indicators is bleak.

Effective utilisation of space

Effective utilisation of space is frequently quite high on the agenda for measuring facility efficiency. There is a trap in that managers can interpret effective use of space as the attainment of a maximum density of usage: how much that can be packed into a building, including people.

For an organisation operating with a policy of high churn rates in order to support organic growth, say, or of fluid re-distribution of personnel to support flexible work processes, the effective utilisation of space may require fallow space to sustain churn and daily change. Therefore, effective space utilisation may involve providing volatile departments with a measured surplus of space to allow them to accommodate flexibility in staff without recourse to disruptive relocational churn. Note then that space utilisation indicators, which may seem to be straightforward and quantitatively measurable, could hide many facets.

One organisation which I have done research with used to run an 85 per cent churn rate on a staff size approaching 6,000 and churning across a portfolio of more than twenty buildings. There was no charging of direct churn costs, but the consequential costs of churn, such as lost productivity, can be punitive, and it is not unusual to hear figures quoted such as USD 4,000 per person. How do the figures illuminate the value/burden balance to business processes of a high rate of churn?

Management of maintenance

In one organisation that has been involved in our research, we spoke with business managers as well as facility managers to find their agreed set of indicators. The core business managers were uninterested in considering any facets of maintenance management performance below an aggregated level of indicator for maintenance. Their attitude was that details simply did not enter their considerations; if it's leaking, we'll tell you, and then you fix it. Thus, the management of maintenance was the single indicator reference to facility maintenance in their portfolio of indicators. When activated it meant problems, when quiescent (i.e., managed well) its value invisible. However, this attitude can also lead to difficulties when negotiating flexibility in budgets, and probably also to a disinclination to invest in preventative maintenance. It also meant that however excellent the maintenance regime was, the attainment of the target level of simply satisfactory was all that facility management were going to get recognition for, and whether the target had been met would be assessed by facility users only on the basis of absence of leakages or other manifestations of maintenance breakdown.

Value for money

Value for money is an instructive example of a highly aggregated performance indicator. Trying to define value for money requires a range of other indicators to be defined, measured and evaluated, and then a further round of comparative analysis to be made. This can make the use of highly aggregated indicators unhelpful as a base for decisions. Value for money in the context of facilities management depends on quality of service as well as price or cost—value products may be low quality and low price, or high quality or moderate price. Without knowing a fuller context, it is very difficult to assess value for money.

8.5 MEASURING THE FM CONTRIBUTION TO BUSINESS

Generally speaking, there are a few principles that should be followed when designing a set of performance indicators, regardless of the particular context. Slater *et al.* (1997) recommend that the number of performance indicators should be minimised and that a performance assessment framework should be limited to between seven and twelve performance measures. They also stress that it is essential to have a thorough understanding of the relationships between key business processes and performance measures. Vokurka and Fliedner (1995) have identified non-financial qualitative performance attributes that could support the holistic and strategically oriented modelling of critical success factors. They have emphasised that a balance between financial and non-financial measures is desirable, since no single measure can adequately provide a clear performance target.

Just as organisational performance can be measured by the value the organisation adds to the inputs that it consumes (Kay, 1995), we should try to measure the internally added value that any support function provides to an organisation. Looking first to the uppermost level of facilities management performance, where the measures should indicate the added value that facilities management provides to the business, Varcoe (1994) has made a strong case for selecting performance indicators related to cost, quality and delivery.

Varcoe also noted the inherent limitations of using financial measures such as cost-per-capita or costs-per-area. In his 1996 paper, Varcoe went on to discuss the relevance of careful selection of performance indicators that resonate with core business performance indicators, in order to reflect the business contribution of facilities management (he looked at a manufacturing context). Varcoe also observed the inherent limitations of using financial measures alone, such as cost-per-capita or costs-per-area. He also suggested limiting the number of performance indicators, and that it is usually sufficient to have five or six well-defined business objectives, each with four to six key facilities performance indicators. The larger the number of indicators the more difficult it is to keep track, and more indicators doesn't necessarily mean a better service or better monitoring of it.

8.6 CONCLUSIONS

In 1990, Frank Becker observed that almost no organization is able to provide at a glance a kind of facilities management balance sheet that shows the relation of these individual indicators to one another. This is a situation that remains today, and thus we still need the basic research before performance measurement practice in the field of facilities management can advance. Meanwhile, facilities management runs the risk of defining what it does by what it measures, rather than measuring what it contributes.

An integrated solution requires an integrated approach, and joint development of business-specific key performance indicators would help. But integrated development of indicators resolves the issue only locally within a single organisation, and it also requires bravery to step aside from a focus on external benchmarking. Business value arising from facilities management also varies for the same organisation when it faces different market environments; it is a tenet of strategy that a firm must be able to match its strengths with the opportunities of the environment (Chakravarthy, 1986). Yet it is beyond this even that lies the realm of facilities management attuning workplace provisions to business benefit—how to optimise the facilities management spend/business benefit ratio for different business processes, or the same process in different market circumstances (Hinks, 2000a). If we assume that performance indicators are to be used within a framework of continuous improvement, we also have to acknowledge that also indicators will have to change over time (Varcoe, 2002).

Continuing the hunt for a Silver Bullet KPI, and of measuring and comparing facilities management using lowest common denominators that are data-driven rather than needs-driven, and hoping that the baggage that the data carry can safely be overlooked, is no solution. Plus, as Tsang, Jardine, and Kolodny (1999) noted, what is needed also is research into soft measures such as the fit between organisational culture and the structuring of maintenance work; the vertical alignment of objectives at different levels of the hierarchy; and the horizontal integration across multiple functions that interact with maintenance. We need the grey data that reflect how complex the integrative role within facilities management is—such information cannot be disregarded because it is subjective and difficult to identify, quantify and measure in analogy with the management of occupational risks (Falconer, 2002). This is where the ability to illuminate the links between facilities management and business performance is more likely to come from.

8.7 REFERENCES

Amaratunga, D., Baldry, D. and Sarshar, M., 2000, Assessment of facilities management performance: what next? *Facilities,* **18**, pp. 66–75.

Ammons, D.N., 1999, A proper mentality for benchmarking. *Public Administration Review,* **59**(2), pp. 105–109.

Baldry, C., 1997, Hard day at the office: the social construction of the workplace. University of Strathclyde Department of Human Resource Management Occasional Paper, 9 February.

Becker, F., 1990, *The total workplace: facilities management and the elastic organization,* (London: Van Nostrand Reinhold).

Bordass, B., Leaman, A. and Ruyssevelt, P., 1999, Get real about building performance: findings from the Probe surveys, and their implications for the procurement and management of buildings. Probe Strategic Review. Report 4: Strategic Conclusions. May.

Carlopio, J.R. and Gardner, D., 1992, Direct and interactive effects of the physical work environment on attitudes. *Environment and Behaviour,* **24,** pp. 579–601.

Chakravarthy, B.S., 1986, Measuring strategic performance. *Strategic Management Journal,* **7,** pp. 437–458.

Davis, G., and Szigeti, F., 1982, Planning and programming offices: determining user requirements. *Environment and Behaviour,* **14,** pp. 302–315.

Falconer, L., 2002, Management decision-making relating to occupational risks: the role of grey data. *Journal of Risk Research,* **5,** pp. 23–33.

Falconer, L. and Hoel, H., 1997, Occupational safety and health: a method to test the collection of grey data by line managers. *Occupational Medicine,* **47,** pp. 81–89.

Hardy, V. and Hinks, J., 2001, Stand and deliver: internal benchmarking and your future in FM. Keynote address at FMA Ideaction 2001, Facilities Management Association of Australia, Melbourne, May 10–11.

Hedges, A., 1982, The open plan office: a systematic investigation of employee reactions to their work environment. *Environment and Behaviour,* **14,** pp. 519–542.

Hinks, J., 2000a, Measuring the value of FM to differing business scenarios. Available www.fmlink.com.au/images.au/Papers/hinks2.htm (accessed 7 May 2003).

Hinks, J., 2000b, Measuring the important, or the importance of measurement? *Facilities Management World,* No. 19 (Summer), pp. 11–13. Available www.fmlink.com.au/images.au/Papers/hinks.htm (accessed 7 May 2003).

Hinks, J., 2000c, FM performance and accountability. In *Facility Management: Risks and Opportunities,* edited by Nutt, B. and McLennan, P., (Oxford: Blackwell Science), pp. 61–70.

Hinks, J. and McNay, P., 1999, The creation of a management-by-variance tool for facilities management performance assessment. *Facilities,* **17,** pp. 31–53.

Hofstede, G., 1981, Management control of public and not-for profit activities. *Accounting, Organizations and Society,* **6,** pp. 193–211.

Kaplan, R.S. and Norton, D.P., 1993, Putting the balanced scorecard to work. *Harvard Business Review,* September/October, pp. 134–147.

Kay, J., 1995, Foundations of corporate success: how business strategies add value. In *Performance Measurement and Evaluation,* edited by Holloway, J., Lewis, J. and Mallory, G., (London: Sage), pp. 280–289.

Khatri, N. and Ng, H.A., 2000, The role of intuition in strategic decision making. *Human Relations,* **53,** pp. 57–86.

Kincaid, D.G., 1994, Measuring performance in facilities management. *Facilities,* **12**(6), pp. 17–20.

Leaman, A., 1995, Dissatisfaction and office productivity. *Facilities,* **13**(2), pp. 13–19.

Leaman, A., Cassels, S. and Bordass, B., 1999, The new workplace: friend or foe? *Environments by Design,* **3**(1).

Leifer, D., 1998, Evaluating user satisfaction: case studies in Australasia. *Facilities,* **16**(5/6), pp. 138–142.

Massheder, K. and Finch, E., 1998, Benchmarking metrics used in UK facilities management. *Facilities,* **16**(5/6), pp. 123–127.

McDougall, G. and Hinks, J., 2000a, Identifying priority issues in facilities management benchmarking. *Facilities,* **18**, pp. 334–337.

McDougall, G. and Hinks, J., 2000b, Exploring the issues for performance assessment in facilities management. In *Proceedings from the CIB W70 International Symposium on Facilities Management and Asset Maintenance, Brisbane, 15–17 November,* pp. 251–258.

McDougall, G., Kelly, J.R., Hinks, J. and Bititci, U.S., 2002, A review of the leading performance measurement tools for assessing buildings. *Journal of Facilities Management,* **1**, pp. 142–153.

Oldham, G.R. and Brass, D.J., 1979, Employee reactions to an open plan office: a naturally occurring quasi-experiment. *Administrative Science Quarterly,* **24**, pp. 267–284.

Payne, T., 2000, *Facilities management: a strategy for success,* (Oxford: Chandos).

Preiser, W.F.E., 1995, Post-Occupancy Evaluation: how to make buildings work better. *Facilities,* **13**(11), pp. 19–28.

Reade, Q., 2001, Benchmark—or be out of a job. *The Facilities Business,* **18**(August), p. 2.

Slater, S., Olson, E.M. and Reddy, V.K., 1997, Strategy-based performance measurement. *Business Horizons,* July/August, pp. 33–44.

Smith, P., 1995, Outcome-related performance indicators and organizational control in the public sector. In *Performance Measurement and Evaluation,* edited by Holloway, J., Lewis, J. and Mallory, G., (London: Sage), pp. 192–216.

Tsang, A.H.C., Jardine, A.K.S. and Kolodny, H., 1999, Measuring maintenance performance: a holistic approach. *International Journal of Operations and Production Management,* **19**, pp. 691–715.

Varcoe, B., 1994, Facilities performance: achieving value for money through performance measurement and benchmarking. *Property Management,* **11**, pp. 301–307.

Varcoe, B., 1996, Business-Driven Facilities Benchmarking. *Facilities,* **14**(3/4), pp. 42–48.

Varcoe, B., 2002, The performance measurement of corporate real estate portfolio management. *Journal of Facilities Management,* **1**, pp. 117–130.

Vliet, A. van de, 1997, Are they being served? *Management Today,* February, pp. 66–69.

Vokurka, R.J. and Fliedner, G., 1995, Measuring operating performance: a specific case study. *Production & Inventory Management Journal,* **36**, (1), pp. 38–43.

Walters, M., 1999, Performance measurement systems: a case study of customer satisfaction. *Facilities,* **17**, pp. 97–104.

CHAPTER 9

Developing Balanced Scorecards for Facilities Management

Dilanthi Amaratunga and David Baldry

9.1 INTRODUCTION

In 1991, Eccles predicted, 'Within the next five years, every organization will have to redesign how it measures its business performance'. Given the current levels of activity in the field, it appears that Eccles' assertion was fair. This paper is about performance measurement of support services for an organisation, more precisely those services usually referred to as facilities management. It covers the application of the Balanced Scorecard which translates an organisation's mission and strategy into a comprehensive set of performance measures that here provide the framework for a strategic measurement and management system for facilities management.

Furthermore, we are able to show how non-financial criteria are as important as financial criteria in measuring the performance of a facilities management function. The implications of Balanced Scorecards for facilities management are reviewed, based on a pilot study carried out in an institution for higher education, and a set of propositions that may form the basis for further research are suggested.

9.1.1 Transferring the scorecard to the context of facilities management

Starting from the four perspectives of the original Business Scorecard as devised and described by Kaplan and Norton (1993; 1996), there are four questions to be asked in the context of a facilities management function. First, there is the customer—how do the facilities users see us? Second, we have to ask related to internal processes—how efficient and effective is the delivery of facilities management services? The third perspective is financial—how is the facilities management function managed in terms of value for money? Fourth, learning and growth—how does the facilities management function continue to improve in itself and to assist the core business?

Given the characteristics of the operating environment of facilities management, recognising and satisfying the needs of the core business is vital for long term commercial survival. To ensure satisfaction of various customer needs, it is essential that facilities management identifies, focuses on, and monitors key performance indicators.

9.2 FM PERFORMANCE IN THE UNIVERSITY SECTOR

This study concentrated on higher education facilities management establishments as the unit of analysis. Property is important to all businesses and organisations. The cost of this asset alone, procuring, managing and operating it, should make it a resource that is high on the agenda of business managers. This applies to all organisations including universities (Housley, 1997). From a business point of view and from a public accountability one, the effective and efficient management and use of the property resource is imperative for all higher education institutions.

The case referred to in this study is a Facilities and Estates Department of a University situated in the Northwest of England.

A detailed review of existing literature on the practice and theory of facilities management with particular emphasis on building performance in higher education, current applications and uses of the Balanced Scorecard in other manufacturing the services industries was conducted. Literature on facilities management in general (Barrett, 1995; Lawrence, 1995; Spedding and Holmes, 1994; Alexander, 1996), literature on building performance (Housley, 1997; London *et al.*, 1995; Preiser *et al.*, 1988) and literature on how to construct Balanced Scorecards (Kaplan and Norton, 1996; 2001; Letza, 1996; Davis, 1996; Walker, 1996) were used to identify important building performance evaluation constructs in the higher educational environment and possible outcomes. Reviewing and evaluating the literature served to identify the key relational concepts that are operational in the given building type including the elements to be dealt with and the criteria to be addressed.

A pilot survey, focused on finding the practical issues of facilities management performance in a higher educational setting. This ultimately helped to uncover the type of information that was required to carry out a more comprehensive survey at the next stage. At the end of this stage, a set of propositions was constructed to cover all necessary relationships among the important issues.

9.2.1 Collecting data

In accordance with the original process outlined by Kaplan and Norton (1996) for developing a Balanced Scorecard, our data collection with the hierarchy of the facilities management organisation and other customers was structured by three main stages. First, we analysed the vision and corresponding objectives of the facilities management function. Second, we identified critical success factors in relation to these objectives, and finally, we developed a set of performance measures to support the critical success factors.

We have taken it for granted that if facilities management organisations are to design fully effective performance measurement systems, it is essential that management can clearly determine what their precise performance measurement information needs are. Barrett (1992) identified certain goals common to most facilities management organisations and this evidence suggests a need for a balanced scorecard approach, which adequately reflects the characteristics, goals and critical success factors of the facilities management organisation.

It was found that the organisation's vision and objectives were to provide a distinctive service combined with value for money, to respond quickly to changes of customers needs, to achieve continually improving services, to develop skills of all employees, and finally to recognise their performance by means of opportunities for advancement. Indeed, these aims present no surprises and they serve to confirm the relative importance of customer satisfaction and achieving good value for money. Obviously, an effective and useful Balanced Scorecard for use by facilities management organisations has to reflect aims such as these.

One of the important issues concerning the development process of the Balanced Scorecard is that all concerned basically agree on the general characteristics of the organisations' important decisions. Therefore, the purpose of the Balanced Scorecard development process is to encourage the discoveries, which are so essential for the development of the organisation as a whole. The quality of the entire process will improve markedly if the participants have been provided with relevant background documentation as well as the opportunity both to question it and to develop it further.

9.2.2 Devising the Balanced Scorecard

The following steps are guidelines for developing the Balanced Scorecard. The process must be tailored to the needs of each organisation. Table 9.1 provides an overview of the process and also indicates the nature of the work.

Interviews were conducted with the senior management to gather data. During interviews, it is important to develop a view of the organisation and its characteristics from as many angles as possible. Critical success factors and appropriate measures were determined, based on the management responses and keeping in mind the vision and the objectives. By focusing on the aspects of the business which created value for customers and, by carefully re-appraising the organisational philosophy and incorporating this into the performance measurement system, this study was able to build a Balanced Scorecard at the exploratory stage, acting as a set of propositions to be tested at the next phase of the study. Tables 9.2 through 9.5 illustrate the critical success factors and corresponding measures identified for each perspective of the Balanced Scorecard.

In emphasising the suitability of the Balanced Scorecard for a facilities management organisation, as a result of our initial, preparatory research, a number of points had arisen which needed to be clarified at the exploratory phase of the study. There appears to be a definite need to be clear about the business unit for which a scorecard is being developed. We certainly envisaged that a scorecard developed for the entire business unit would differ from scorecards for each of the areas or departments controlled by management within the business unit. Moreover, there is a clear need for the Balanced Scorecard components to be reviewed and, where necessary, updated on a regular basis if the scorecard is to remain both relevant and useful. There is also a number of areas that could be used to augment a facilities management scorecard, such as measurements reflecting staff reaction to student needs.

Table 9.1 The process for developing the Balanced Scorecard.

Stage	Task	Deliverables
Establish vision and scope	1. Select appropriate organisation unit and identify its linkage with external bodies	Organisation vision and mission BSC objectives, scope and links with EFQM Appreciation of external objectives Detail plan for next stage
	2. Review EFQM as well as other organisational development agendas 3. Agree organisation vision and mission and BSC objectives	
Establish first draft of BSC	4. Conduct first round of interviews	First draft of BSC, with a list of objectives and critical success factors, plus potential measures for each perspective Detail plan for next stage
	5. Synthesis Session 6. Executive Workshop: First Round	
Complete BSC and its measurements	7. Subgroup Meetings	The BSC Communication strategy to the rest of the organisation Detail plan for next stage
	8. Executive Workshop: Second Round	
Develop implementation plan	9. Develop the Implementation Plan	Final BSC with stretch targets Implementation programme Continuous review programme & infrastructure for BSC
	10. Executive Workshop: Third Round 11. Finalise the Implementation Plan	

Table 9.2 Customer perspective: critical success factors and measures.

Critical success factors	Possible measures/measurement instrument
Customer satisfaction Service quality Customer complaints Range of services offered Reaction to customers' needs	Customer satisfaction surveys post-occupancy evaluation

Table 9.3 Internal processes perspective: critical success factors and measures.

Critical success factors	Possible measures
Service excellence	Service standards, service quality survey
Technology capability	Equipment costs, post-occupancy evaluation
Understand the customers	Customer satisfaction surveys
Employee competence	Employee qualifications, training hours per employee, employee satisfaction index
Process efficiency	Output/cost ratio
Teamwork and co-ordination	Interdependent meetings, interdependent training courses
Staff development	Courses completed, number of multi-skilled staff

Further research in the exploratory phase will address three issues. First, there will be a review of common goals, critical success factors and performance measurement relevant to facilities management. Second, we intend to identify possible Balanced Scorecards appropriate to other levels within the facilities management unit and also consider the implications of using different Balanced Scorecards at different levels in a facilities management organisation. The third issue refers to further development of effective performance measures appropriate to facilities management such as customer satisfaction and staff development, which have been identified as being vital types of measures.

Table 9.4 Financial perspective: critical success factors and measures.

Critical success factors	Possible measures
Management expectations	Cash flow, cost reduction rates, costs per unit of output, new business development
Financial growth	Balance income and expenditure Financial reporting
Cost reduction, productivity improvement	Cost per unit, reduction of indirect costs, services sharing with other business units
Asset utilisation	Reduction of working capital
Management of working capital	Average rate of return

Table 9.5 Learning and growth perspective: critical success factors and measures.

Critical success factors	Possible measures
Technology leadership	Time to develop new processes
Continuous service improvement	Service innovation cycle time, employee turnover, staff attitude survey, number of employee/customer suggestions, development areas identified, new facilities/services introduced
Upgrading staff competencies	Employee satisfaction, staff development programmes, courses completed, internal promotions made

9.3 CONCLUSIONS

This paper addresses the applicability of the Balanced Scorecard into facilities management. The adoption of the process leading to Balanced Scorecards would be a major change initiative in most organisations.

Kaplan and Norton (1996) provided a useful generic model in the form of their Balanced Scorecard. As seen in our study detailed here, the quadripartite model they present is suited to different types of business situations. The original idea of the Balanced Scorecard technique was, as they point out, 'not whether you had created value, but if you are going to create value in he future'.

Facilities management's ability to plan, anticipate and initiate change is enhanced if it utilises management tools such as the Balanced Scorecard. However, in the words of one interviewee from our study, 'the scorecard must be balanced in order to facilitate the achievement of the short, medium and long term goals and objectives of the organisation'. Another interviewee said, 'It sets goals for facilities management and you can bring that back to the table and articulate what facilities management did and how it helped improve the process for those who are not directly engaged with facilities management'.

The study reported in this paper provides the basis for the construction of a set of propositions for a FM Balanced Scorecard, which forms the framework for further study at the next phase. The impact of this management and measurement process is the basis for the research currently being conducted by the authors. Issues such as completeness of the Balanced Scorecard measures, effectiveness of enhancing facilities management performance and barriers to implementation are the focus of the future investigations.

9.4 REFERENCES

Alexander, K., 1996, *Facilities Management: Theory and Practice*, (London: E & F Spon).

Barrett, P., 1992, Development of a Post Occupancy Building Appraisal Model. In *Facilities Management: Research Directions*, edited by Barrett, P. (London: RICS Books).

Barrett, P., 1995, *Facilities Management: Towards Best Practice,* (Oxford: Blackwell Science).

Davis, T.R.V., 1996, Developing an employee balanced scorecard: linking frontline performance to corporate objectives, *Management Decision,* **34**(4), pp. 14–18.

Eccles, R.G., 1991, The performance measurement manifesto, *Harvard Business Review,* January/February, pp. 131–137.

Housley, J., 1997, Managing the estate in higher educational buildings, *Facilities,* **15** (3/4), pp. 72–83.

Kaplan, R.S. and Norton, D.P., 1993, Putting the balanced scorecard to work, *Harvard Business Review*, September/October, pp. 134–147.

Kaplan, R.S. and Norton, D.P., 1996, *The Balanced Scorecard: Translating Strategy into Action,* (Boston, MA: Harvard Business School Press).

Kaplan, R.S. and Norton, D.P., 2001, *The Strategy-Focused Organization: How Balanced Scorecard Companies Thrive in the New Business Environment,* (Boston, MA: Harvard Business School Press).

Lawrence, D., 1995, Facilities Management: Case studies in UK, Europe, North America. In *A Focus for Building Surveying Research*, (London: RICS).

Letza, S.R., 1996, The design and implementation of the balanced business score card: an analysis of three companies in practice, *Business Process Re-engineering & Management Journal*, **2**(3), pp. 54–76.

London, K.A., Chen, S.E. and McGeorge D., 1995, An integrated building evaluation/cost approach that considers stakeholders objectives. In *Proceedings of the International Conference on Financial Management of Property and Construction,* Newcastle, N. Ireland, pp. 456–461.

Preiser, W.F.E., Rabinowitz, H.Z. and White, E.T., 1988, *Post-Occupancy Evaluation*, (New York: Van Nostrand Reinhold).

Spedding, A. and Holmes, R., 1994, Facilities management. In *CIOB Handbook of Facilities Management*, edited by Spedding, A., (Harlow: Longman Scientific & Technical), pp. 1–8.

Walker, K., 1996, Corporate performance reported revisited – the balanced scorecard and dynamic management reporting, *Industry Management and Data Systems*, **96**(3), pp. 24–30.

PART FOUR

Towards Knowledge Workplaces: An Introduction

Tore I. Haugen

The future facilities manager is part of a knowledge economy and society, with rapid changes in business, organisations and work patterns. The traditional workplaces and organisational structure are changing, and we have to rethink the traditional workplaces like offices, schools and industrial facilities. The changes are driven and enabled by the information and communication technology (ICT), giving us the opportunity to work remotely from home or from the train. At all times we might be connected to our networks, as well as we are also at all times accessible for our business. This changes our work pattern and workplaces, moving us from a typically dedicated and function-based workplace to a free address in a satellite office running on a 24-hour basis.

ICT is also changing the way we are working, enabling us to have information and communication systems which give all workers in company, in one building or in many buildings and locations, individual access to all the common knowledge in the company. This knowledge built up from individual tacit and explicit knowledge, individual experiences and common shared experiences are the main values in organisations and networks that are knowledge based, knowledge intensive and knowledge producing.

From a facilities management point of view we have to ask how we can create good workplaces for knowledge work taking into account organisational issues and relations, ICT and digital frameworks and the physical frameworks and architectural design. We have to create settings and systems that enable communication and interaction between the staff, promote creativity and support efficiency in work processes. Improving knowledge sharing and training is important as well as to attract, develop and retain skilled employees. This means attending to staff satisfaction and the creation and management of a good working environment.

From a traditional facilities management perspective, the focus on knowledge workplaces and good settings for knowledge producing organisations is rather new, and we have to ask some basic questions to understand what are the new challenges: What do we understand by knowledge and knowledge work? What kind of literacy, the ability to understand and use information, is needed to function effectively in a knowledge organisation? What are the implications regarding space requirements seen from an individual point of view, from the team or from the organisation? Can we develop structures and typologies that give space solutions to different knowledge organisations? Are there specific and new challenges for facilities management in creating the knowledge workplace?

The understanding of knowledge and knowledge organisations (*k*-organisations) is an important part of the work undertaken by Linariza Haron, who starts Chapter 10 with a summary of the policy options of space allocations as part of our modern ICT-driven society with management responses like the 24 hours office, shared facilities and multi-use functions. Moving on from this perspective, Haron guides us through the basic understanding of the essential elements of knowledge, differentiating between tacit and explicit knowledge, and knowledge work activities at the individual, team and organisational level. In knowledge work, the access to information and the exchange of knowledge are crucial, and that requires the knowledge workers to be highly literate. But what kind of literacy does the knowledge workplace demand? Haron explores the different kinds of knowledge work and literacy, and concludes that since the nature of knowledge work is largely collaborative, workplaces should provide flexible and intuitive support for multi-tasking work styles, not the traditional practice based on rank and status.

In Chapter 11, Reidar Gjersvik and Siri H. Blakstad ask if the new ways of designing workplaces are based on knowledge and understanding of how knowledge-intensive organisations really work and how this work can be best supported—or if the new ways of designing offices are based on a superficial understanding and on loose theories for knowledge work. In their contribution, they describe a range of knowledge work that takes into account the nature of the work and organisations, rather than in trends in office design and management. They propose the use of knowledge work archetypes to describe various recognisable forms of knowledge work, and how this may be related to a typology of knowledge workspaces. Knowledge work archetypes are intended as tools that the user organisations can rely on for developing their workplace requirements. In the second part of their chapter, Gjersvik and Blakstad describe a framework for defining a typology of knowledge workplaces, and they suggest that work archetypes should be linked to a spatial typology based on the user organisations' needs and requirements. Their work is part of a larger ongoing research project developing practical guidelines and solutions together with different organisations and users.

Our knowledge about the knowledge organisation seen from a design and management viewpoint is growing, but we are still in the early stages of understanding what really matters in a knowledge workplace. We have to develop concepts and methods further in order to understand and describe knowledge intensive work, to understand and describe what physical spaces and places do for the knowledge work and the role of the management and control for the knowledge workplace. These issues have been one focus for the EuroFM research network in the last years, and will be an important research issue for the coming years. We still have a long way to go to see the full implications for the facilities manager.

CHAPTER 10

The Knowledge Workplace: What Really Matters

Linariza Haron

10.1 INTRODUCTION

Knowledge work, an essential activity in the new economy, whether we believe in it or not, driven and enabled by information and communications technologies (ICT), has resulted in drastic changes in work organisation and in workplace provisions. Various workplace arrangements, space support and utilisation strategies have accommodated these changes. But are these strategies relevant for the knowledge workplace? We conducted an extensive review of literature on workplaces and concentrated our search to workplace literacy. We begin to understand that knowledge workplaces must support the different modes of communication in knowledge exchange, inter- and intra-organisationally, at all operating levels through the full range of space allocation options, from dedicated assignments to free address allocation. This guidance should assist facilities management in planning for workplaces so as to fill the spectrum of diversity requirements among knowledge workers.

Studies abound explaining the radical restructuring of the economy. The new economy has been said to be synonymous with the new service economy, the information economy, and the knowledge economy, indicating the growth of service predominant or information processing jobs (Castells and Aoyama, 1994; Roberts *et al.*, 2000), supported by ICT as its technology base (Howells, 2000, p. 272). Knowledge, rather than physical facilities, is increasingly defining a competitive advantage (Bassi *et al.*, 1998) as services take on a more facilitating and partnership role with other parts of the economy (Howells, 2000, p. 275).

Organisationally, the processes and products aim at accomplishing global business goals, in achieving cost efficiency, greater productivity and increased company profits. Structurally, knowledge organisations tend to be in network arrangements (Birchall and Lyons, 1995, p. 88) and emerging out of new concepts and forms (Miles and Snow, 1986).

These knowledge organisations (or *k*-organisations) are present across most, if not all, industrial economic sectors, and as long as 'advanced technologies are employed, all industries can be knowledge intensive' (Porter, 1999). Their occupational functions are also broadly affected, particularly functions that handle the processing, assimilation and synthesis of information, to acquisition, creation, packaging, distributing, applying and maintaining of knowledge; which differentiate the knowledge workers from the non-knowledge workers. Knowledge workers keep the firm abreast of developments, business opportunities or risks available; perform as internal consultants; engage in problem solving; and act as agents of change (Mukerji, 2000, pp. 185–186).

Yet, the drivers to organisational changes are wide-ranging. Human resources, operational and financial are some common reasons for such changes (Avis and Gibson, 1996). A *k*-organisation is focused on creating and disseminating knowledge and is still profit oriented. The enablers are the people, process, workplace and IT (Davis *et al.*, 1985; Birchall and Lyons, 1995; Eclipse Group, 1995; Laing, 1993). The first represents the human capital, the second emulates the intelligence, the third facilitates spatially and the fourth supports technologically. In some instances, the third and fourth elements are integrated. Therefore under contemporary business conditions, where there are fundamental changes as to how work is organised and who performs it within space and time, requirements for work settings that are flexible to accommodate the fluid pattern of work (Joroff *et al.*, 1993, p. 84) are becoming vital.

This chapter explores some essential attributes of a knowledge workplace provided for the knowledge workers who actually do the knowledge work and are themselves the users. We define the knowledge workplace as the support to information transfer and knowledge exchange through various communication strategies at all operational levels in the organisation. Our objective is to develop guidelines for space allocation, linking knowledge exchange activities and their modes of communication. We anticipate that these guidelines would assist the management of facilities in planning knowledge workplaces that are flexible and accommodate the diversity of requirements of knowledge workers.

10.2 THE WORKPLACE

A workspace is the individual area that people occupy (Apgar, 1993), whereas the workplace is an enabler of work processes (Becker and Joroff, 1995, p. 6). The two other descriptors are their working arrangements and facilities landscape (Price, 1997). The Facilities Management literature reports these under the strategies of the changing workplace, whether they are organisational and employment, space support or utilisation types. We review these sequentially.

10.2.1 Organisational and employment strategies

Forms of workplace arrangements vary according to types of employment strategies. Contractual working arrangements fundamentally differentiate the legal status of workers on the grounds of contract *for* service against a contract *of* service (Bertin and Denbigh, 1996, p. 140). The former identifies the self-employed while the latter could refer to part-timers, contract workers or teleworkers. Their legal status directly affects physical facilities provision in the workplace.

Similarly, organisational strategies through restructuring has seen tasks being performed and communicated throughout the entire work process through team working, partnership or consultancies. This, together with the related employment strategies, has created a new work culture, which in turn has changed the workplace from being a real estate commodity to a set of resources that could be allocated and managed in a dynamic way (Joroff *et al.*, 1993, p. 85). The transformation from physical entities like head offices, satellite offices, business centres, telecentres and telecottages into mobile virtual arrangements via the laptop or cell phones (Birchall and Lyons, 1995, p. 111; Becker and Joroff, 1995, pp. 6–7; Zuboff, 1988, p. 387) also affects how geographical locations are chosen (Apgar, 1993).

10.2.2 Space support strategies

Within the range of space support strategies, we have distinguished planning and allocation, workplace design / IT support and utilisation strategies. Planning and allocation are necessarily taken together, as the procedure for allocation is normally based on a space plan, which programs the workspace allocation, working group needs, group adjacencies and departmental groupings within the organisational strategies: growth, steady state or reduction (Zeisel and Maxwell, 1993, p. 163). Table 10.1 shows the range of allocation options available, from dedicated space assignments, space sharing through to independent workplace types for changing organisations.

We have identified two different perspectives for the workplace design / IT support strategies. First, the spatial features of existing buildings are found to be no longer compatible with the changing work organisations of the business world. There are constraints to the capacity, configuration, structure and dimensions of the available workspace that subsequently affect the occupancy levels, utilisation, subdivision and the layout arrangements of the facilities (Haron, 2000). In this situation, adaptability measures and systems furniture arrangements are among the common responses.

Second, the activities and ways of working of knowledge workers, are distinctly different from the traditional; they can be directly associated with particular space types. Hence, an individual's work process is being allocated a 'hive'; concentrated study into 'cell'; group process into 'den' and transactional knowledge into 'club' (Laing, 1996). One of the earliest works of this type of space support strategy was that of Stone and Luchetti (1985) followed by the Integrated Workplace Strategies, Found Alternative Workplace, Design Alternative Workplace (Offices), Team Offices, Cyberspace and Virtual Organisations (Becker and Joroff, 1995).

10.2.3 Workplace utilisation strategies

Utilisation strategies, however, arise from the erratic pattern of space utilisation. The 'true' use of space is related to the activities that people engage in over time, number of occupants and the duration of use. Studies carried out at the building level claim that facilities are utilised between 5 per cent (Lloyd, 1993) to 15 per cent of the time for which they are available (Varcoe, 1995; Chadwick, 1993; Digital, 1993).

Therefore, it is reasonable that one of the prevailing goals in the changing workplace is to optimise space utilisation. Major studies in this area have already been carried out for educational facilities since the 1960s. The business sector has only recently adopted this approach. Some typical management responses are the 24 hours office, shared facilities, booking systems, clear desk policy, multi-use functions, charge back system, and the implementation of increased availability of space hours.

Table 10.1 Summary policy options via space allocation.

Allocation procedure	Options	Description and influence on management
Typically dedicated	Function based	The user's job requirements determine the amount and type of space and the furniture provided. This can result in a complex inventory of standards and furnishing components associated with each type of department, group and individual user.
From dedicated to space assigned	Zero based	The provision of workspace and furniture is negotiated on an individual basis, based user's stated requirements. Senior executives often expect this.
From dedicated to space sharing	Universal plan/footprint	A workstation of the same size is provided for all space users. It results in low churn costs, as people, rather than panels, are moved.
From space assigned to space sharing	Non-territorial	The broad concept is that space and furniture are viewed as a company asset for individual use but only as and when required. Variants within this type of policy include shared space, hotelling, team base and home base. As a space allocation practice in which individuals are not assigned desks, workstations or office for their exclusive use.
From dedicated to free address	Satellite office	As work centres, placed in residential areas or rural villages, but usually owned by a specific company which has relocated part of its operations at a distance from the main site. A form of geographic disposing of employees, and telecommuting from home based office.

Source: Adapted from Watts (1994).

10.2.4 The implications

We have now seen how workplaces have coped with the generally changing organisations. Where *k*-organisations are concerned, the crucial elements relate to the levels of operations and the communication modes and with fundamental change affecting the process base. The link between knowledge exchange and knowledge workplace needs to be examined more thoroughly. We think that the space support strategies mentioned earlier could be further enhanced to better fit the needs of the knowledge workplace. Already Birchall and Lyons (1995, p. 180) suggested better use of technology such as by teleconference centres, cyberfora, multimedia research areas equipped with Internet research and accessing CD ROM based materials, docking stations and flexible database management systems. After all, in the years ahead, workplaces will be electronically linked worksites facilitated by computerised coaching and monitoring equipment. So there is a need to create a compelling workplace that will satisfy the workforce in the *k*-organisation. We should rationalise these arrangements through allocation procedures that support the knowledge work to be done.

10.3 KNOWLEDGE WORK

According to the American Bureau of Labor Statistics in 1998, as referred to by Mukerji (2000), knowledge work has four characteristics. First, there is the existence of a valid codified body of knowledge found in books. Second, the body of knowledge must be capable of being taught; it differs from a skill that can only be learned through experience and apprenticeship. Third, practitioners of the body of knowledge must normally prove their mastery of that knowledge by being certified. Fourth, the professions must maintain [high] standards of admission for the practitioners through regulation by independent professional organisations (Mukerji, 2000, p. 184). Hence, knowledge work is characterized by particular formal arrangements. However, this definition does not explain the subtleties of knowledge work processes.

We are interested in the process that generates new knowledge, and the kind of support that contributes to its success. So we rely on the classification of knowledge into tacit and explicit categories as suggested by Polanyi, also discussed by Howells and Roberts (2000, p. 251). Thus, tacit knowledge is accumulated through experience such as learning by doing, and social interactions, while explicit knowledge is accumulated through professional skills producing knowledge that can be written down in the form of document, manual, blueprint or operating procedures. Using these ideas, we may attempt to understand knowledge creation and the knowledge enabling processes within knowledge work activities.

10.3.1 Knowledge exchange

The processes that generate the creation and the effective transfer of knowledge have already been identified by Nonaka and Takeuchi (1995), to which should be added what Egbu *et al.* (2000) and Blackler (1995) have written. For our purposes, Nonaka and Takeuchi's model helps to illustrate the knowledge exchange process occurring between tacit and explicit categories, as displayed in Table 10.2.

Table 10.2 Knowledge exchange, after Nonaka and Takeuchi (1995).

TO FROM	TACIT	EXPLICIT
TACIT	Socialisation; where people are involved in creative dialogue and the spreading of ideas.	Externalisation of knowledge in written form, using explanatory techniques, such as metaphors to enhance understanding.
EXPLICIT	Internalisation of knowledge, i.e. the embodiment of explicit knowledge.	Combination which involves systematising knowledge for richer understanding and to stimulate new ways of thinking about things.

Source: Adapted from Egbu *et al.* (2000).

The typology of Table 10.2, which is instrumental to our study, shows the range of knowledge activities that occur within a *k*-organisation. This helps in conceptualising the implications of knowledge activities at each operational level: from the individual, the team, to the whole organisation. Individuals tend to do a lot of thinking and planning, working with databases, computer programming, graphics, layout, and are often absorbed in personal reflections. Teams tend to engage in interactions and collaborative work, supported by their project environment through teamwork. The accumulated results are manifested at the organisational level, where members learn collectively from the range of impulses by translating the many cues into appropriate actions (Birchall and Lyons, 1995, p. 162).

Therefore, within the *k*-organisation, the content of work usually requires a high level of cognitive skills focusing on collaborative problem solving as the work is no longer dominated by repetitive operations (Price, 1997). Knowledge is the accumulating result of experience in a form that is shareable. This accumulation occurs as feedback, as information is used and then passed on to others (Winslow and Bramer, 1994, p. 40).

10.3.2 Information literacy

Obviously, it is the access to information and the exchange of knowledge that are crucial in knowledge work, and that requires knowledge workers to be highly literate. Literacy is the ability to understand information and to use it to function effectively. Being literate entails both the skill of deciphering written materials—encoding and decoding—and extracting meaning (Skagen, 1986). But what kind of literacy does the knowledge workplace demand? Do we mean computer literacy? Internet literacy? Multimedia literacy? Media literacy? Network literacy?

The Association of College and Research Libraries suggests that as the workplace goes through 'cataclysmic changes', very few workers will be prepared to participate successfully and productively unless they are information literate (ACRL, 1998). Information literacy is essential in the knowledge workplace. It is 'the ability to locate, process, and use information effectively; [and to] equip individuals to take advantage of the opportunities inherent in the global information society'; a notion expressed by the Association for Supervision and Curriculum Development, stated in a 1991 resolution, and quoted by McClure (1997) from Breivik (1992). McClure also suggested that literacy be seen in the context of information problem-solving skills—the Big Six Skills (McClure, 1997, p. 420), or within the context of the traditional, technical (in computers and telecommunication), media and network literacy. Its relevance to business organisations would ensure that literate professionals could carry out problem-solving tasks, while managers could focus on issues of control and make decisions effectively.

Another important element lies more precisely in the individual's functional literacy. These individuals should possess the ability to obtain information they want and use that information for their own and others' well-being (Skagen, 1986). Hence, they should develop their skill inventories for specific functions to serve as a basis for workplace literacy plans—i.e. their job descriptions. Their literacy facilitates access to high-value content, a tactical intelligence that they can use in their decision-making at their point of need (Corcoran, 2001). Examples of application are in business finance, communications, workplace behaviour and teamwork in general.

However, reports on workplace information literacy initiatives are inadequate. The themes of those that are available are disparate. For example, one study has focused on developing the profile of an information literate law firm, and another on the experiences of auditors and engineers in a workplace. Of the first, Bruce (1999) reported the work of Gasteen and O'Sullivan who had found that information literacy 'on an organisational level, impacts on its success in the market place; [and it] revolves around the library, [...] their Infobank database, as well as human resources and training'. The second study is related to the work of Bonnie Cheuk, who had found that the experiences of such professionals in the workplace are recursive; that information seeking is not always necessary; often working by trial and error; not 'getting the answer'; not linear; not a one man job; and that their relevance criteria change (Bruce, 1999).

10.4 KNOWLEDGE WORKPLACE

It is the qualitative aspect of the knowledge workplace that is emphasized here, rather than the quantitative rate of knowledge exchange and information transfer within the workplace. To do this we have adopted workplace literacy as a proxy for knowledge exchange, simply assuming that workplace literacy indicates the performance of the knowledge workplace. Indicative measures may be gathered from training programmes and systems installed. The more regulated and sophisticated the training programmes and systems installed, the higher the level of workplace literacy achieved, and conversely. The next section will further describe workplace literacy followed by an overview of literacy surveys. After that, we attempt to clarify the two divergent methods of space allocation toward realising the knowledge workplace for knowledge workers.

10.4.1 Workplace literacy

According to Skagen (1986), there are four principles that help define literacy in the workplace. Firstly, need is fundamental; secondly, change in the nature of work has put new demands on the literacy of its workforce; thirdly, each workplace requires its own definition of literacy; and fourthly, the lower operational levels are usually 'pulled up' or assisted by the more literate ones.

All these principles are relevant to *k*-organisations. A partial list has been made available by Imel (1995). Greater emphasis is duly placed on technology and communications, both being vital to the knowledge activities of the individuals, the teams and the organisation. An early study found fifty per cent of the communication in the workplace to be oral (including telephones), thirty-five percent in written form and fifteen per cent as using electronics via computers (Strassman, 1983). In *k*-organisations, particularly in 'the circumstance where people are meeting more often, face-to-face (in same place), or electronically (in different place)' (Bradley and Woodling, 1999), the more advanced communication strategies are already endemic and intense.

Where employees are allowed to exercise their choice of workplace, at least to a certain extent, they are taught as to how to work effectively in the new ways. As more resources are deployed to training and education, in order to build information competencies throughout the workforce (Corcoran, 2001), workers may readily discern, sort, evaluate, and apply vast amounts of information, thus reducing the risk to organisations exposed to a history of low productivity, accidents, absenteeism, and poor product quality. Many organisations are now also committed to providing standardized computer software for people working in all locations, to ease accessibility to qualified technical assistance and financial resources (Apgar, 1998). At best, these efforts will deter employees from clinging to familiar patterns of work practice and instead readily improve on the new, alternative ways of working.

In our effort to understand workplace literacy in *k*-organisations, we have looked closer at a number of programmes and systems that help stimulate knowledge exchange in the workplace. The programmes are categorised either as 'soft' or 'hard' methods.

Knowledge exchange programmes

Among soft methods like mentoring, buddy systems and community practice, the first has been established more successfully in some organisations. In their research, Swap *et al.* (2001) have investigated the use of mentoring and storytelling to transfer and ultimately leverage the knowledge in an organisation. They reviewed management and cognitive psychology literature, finding that the knowledge-sharing mechanism of the approach could be enhanced by informed managers who could understand the informal manner of transfer through the process of socialisation and internalisation. They also found that telling true organisational stories are considered powerful conveyors of tacit knowledge (Swap *et al.*, 2001) compared to the artificially created ones. Mentoring however, takes time and continuity.

Among the hard methods applied, training, education and systems models are forerunners. The change from the industrial age to the information age, from domestic to global competition, from initially simple technology base to the highly sophisticated technology of today all require upgrading of employee skills according to Gordon (1989), as referred by Lankard (1991). Comprehensive training programmes go beyond initial training. Lifelong learning experience is nurtured through work-related learning activities and workplace training.

Lankard also underlines that through these learning systems, employers and employees are able to respond to changes in an efficient and timely manner (*cf.* Carnevale, 1989). For the employer, training supports organisational culture and goals while encouraging efficiency, innovation and quality in worker performance. For the employees, the benefits are both economic and educational. Brown (1998) has discussed what Mosca (1997) claims, namely that they will be infused with creativity into their jobs, will be able to tolerate ambiguity, and accept responsibility and accountability for their work. A case in point is how IT education has progressed from mainframe oriented courses of the 1970s, to microcomputer oriented courses of the 1980s, further to microcomputer based application programmes as productivity tools and more recently, to the use of networks to share resources (Quarstein *et al.*, 1994). Training, in the newest development, involves workshops on the protocols of using e-mail for keeping in touch with co-workers, as well as teaching staff to use upgraded and new software and hardware. End-users are also taught how to structure the workday, prevent burn-out, and use technology to find answers and solve problems when working remotely (Becker and Joroff, 1995, p. 79).

Literacy survey

Table 10.3 shows a sample of workplace literacy programmes and systems across the industry. It shows to what extent knowledge-based organisations prepare the workforce in applying technological innovations to improve business solutions (Thompson, 1997) through their workplace as a support.

This overview of literacy studies shows that although programmes may have diverse aims and objectives, their agenda help the end users to relate to IT and other media; to acclimatise to the realities of home working; to realise the importance of active experimentation and to improve networking skills. There is a progression from basic to advanced training, exercises on rituals, sharing and interrogating databases. All these activities ultimately lead to gains in knowledge. Their target trainees are relocated employees, home workers, but also managers, directors, and partners.

In comparison, Table 10.4 shows a sample of the training and development framework of the traditional non-ICT mediated workplace, which adopts some general and tailor-made programmes.

Table 10.3 Knowledge-based ICT mediated workplace.

Author/ source	Company	System/tools	Effects on workplace literacy
GMR, 2001	MCI Inc.	Online human resources	Employees able to register for training courses online.
Spence, 1999	Labour union partnership with management and others	Labour sponsored worker-centred learning programs	Lifelong training to update technical skills and develop problem-solving skills. Programmes deal with academic, personal management, teamwork skills.
McDermott, 1999	British Petroleum with Shell Oil	Networking people	Thinking together – ability to leverage knowledge as key to its competitive edge.
Apgar, 1998	Merrill Lynch	Telecommuting lab	Acclimatising candidates for the alternative workplace.
Apgar, 1998	AT&T	Survival training	Teaching new norms to alternative workplace.
Apgar, 1998	Lucent	Rituals	Linking the traditional office to new realities of home office.
The Eclipse Group, 1995	BT Westside	Workstyle 2000, Office Automation Optimisation	How to make the best use of IT; from basic to complete redesign of work and/or work practice.
Winslow and Bramer, 1994	Sara Lee Knit	Workplace Literacy System project	Dealing with the real needs of workers and creating an environment in which learning takes place continually.
Skyrme, 1994	Digital	Active experimentation and learning	Learning with the help of external advisors and facilitators, and avoiding 're-inventing the wheel.'
Joroff *et al.*, 1993	R & D labs	Courses in technology and management or work in new form	How to work in different locations, organise time, use new technology, sense problems and resolve them early.
Joroff *et al.*, 1993	Nippon T&T	Remote video training modules	How employees in satellite offices could access video training library, and real-time interactive training.
Joroff *et al.*, 1993	NYNEX	NYNEX shuttle experimental system	Allowing users make real time video connections, sharing written and graphic information with other sites.
Skagen, 1986	Polaroid Company	Technology readiness curriculum	Focusing on skills people need to improve their job performance or to prepare for growth. Training includes components of maths and science, computers and instrumentation and skills for sustained learning.

Table 10.4 Traditional non-ICT mediated workplace.

Author/source	Company	System/tools	Effects on workplace literacy
Skagen, 1986	Onan Corp.	Manufacturing Education programme	Two level curriculum for training: Level I course, to hourly workers and supervisors, no pre-requisites and no job-reassignment. Level II course is job-related and could potentially lead to a different job.
Skagen, 1986	NYNEX	Developmental studies and The developmental skills program	The first is geared to improving basic skills such as effective communication, basic math, and basic electronics. The second, has courses designed to develop the managerial skills of employees.
Ballinger and Gee, 1996	Partnership between College of Lake country and six industries	Workplace literacy program	Standard curriculum design, and development of procedures and resources for instructors. Topics covered: learning styles and strategies, instructional methods, cross-cultural communication, communication skills, problem solving in the workplace. Content specific to the industries involved with general job categories, e.g. machine operators, mechanic, machinist and shipper/receiver.

Here, in the traditional arrangement, programmes may be described as 'a comprehensive, multifaceted curriculum designed to deliver job training and placement services for the underskilled, undereducated, and unemployed' (Skagen, 1986) or that 'these efforts typically concentrate on improving adult reading skills; promoting math and science education; and job training' (McClure, 1997). What was missing was the need for 'other services' (Skagen, 1986) where, in the *k*-organisation, the continuous training exploits the potential that the new hardware and software represent.

Therefore, our overview shows that training the employees will certainly develop competencies and broadly define their employability skills; whether in basic, communication, adaptability, developmental, group effectiveness or influencing skills. These skills could eventually contribute to optimal learning according to Bailey, as mentioned by Overtoom (2000). Over time, the individuals could progress to an expert status. At the organisational level, where companies have developed sufficient competencies, they will flourish as they are endowed with differentiating specialities. But where companies have no specialities, they will have to consider outsourcing arrangements.

10.4.2 Methods of space allocation

We will now explain as to how to manage the whole process of knowledge creation in the context of knowledge space within the existing workplace resources. We suggest that appropriate space allocation procedures be applied when considering the interaction between people and information. Through these procedures, knowledge work is better supported to build an effective and efficient transmission channel and convince the individuals to use the knowledge received. We present two divergent methods: the allocation of activities to space, and vice versa.

Activities to space

Traditionally, the planning of office space has been primarily driven by the allocation of space to individuals, based on their job profiles, communication needs and other functional requirements. Space allocation necessarily operates by activity functions. But in practice, the allocation criteria commonly relate to the status of the individual worker (Langdon and Keighley, 1964; Duffy, 1992) with a significant difference between the allocation of space to higher ranking individuals as compared to lower ranked staff. This has been dominant in office organisations of the past.

Moreover, the space planning at the working group level is organised around individuals who share a similar rank and task and who can therefore occupy the same room. At the departmental level, it is typically organised on a floor by floor or zone by zone basis, depending on the size of the department. At the organisational level, the typical space policy will determine activity functions and their corresponding space types. The layout options could be enclosed cellular offices, open plan layouts and group space types (Index to Coverage of Offices, 1964; Worthington, 1982) or other variations like 'bullpen' plans, multi-occupancy and single-occupancy (Watts, 1994). However, these classifications do not radically change the allocation procedure for individuals and groups.

For *k*-organisations, innovative layout options have direct implications on allocation procedures. Figure 10.1 shows the conceptual shift seen in a continuum illustrating the spectrum of allocation options available, from space assignments (dedicated workstations) and space sharing (shared assigned or space sharing) to a complete independence of space/location (free address satellite offices, or telecentres).

Hence different types of knowledge exchange are effectively supported by their workspace/place types via the allocation system, as shown in Table 10.5. The determining factor would be the types of knowledge exchange involved and their communication strategies. As per allocation, the options selected will differentiate their relevance to knowledge workplaces. The main consideration is that through this allocation procedure, all points of demand are considered in context. As such, what is extremely critical is the constant availability of these places across the organisational facilities.

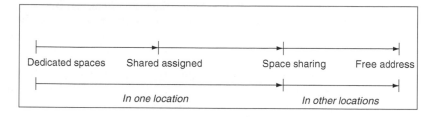

Figure 10.1 The continuum of space allocation procedures for knowledge workplaces.

Table 10.5 Knowledge exchange and the allocation of spaces.

Knowledge exchange	Examples of soft/hard transfer	Communic- ation modes	Allocation	Examples of space types
Tacit to tacit	Mentoring and storytelling	Face-to-face	Space sharing	Group spaces/hub
Tacit to explicit	Collaborative	Text mediated	Free address	IT support universal workstations
Explicit to explicit	Training education	Face-to-face, process oriented	Free address/ space sharing	Learning centres
Explicit to tacit	Mentoring and storytelling	Personal self- reflection	Dedicated space/shared assigned	Quiet space/safe haven

Space to activities

On the other hand, most physical facilities are unable to transform as fast as organisations change. Given this lack of flexibility, we have to allocate available space to activities. In the past, when a building was occupied, the basic management question was: on what basis could the available space be better allocated, to support the activities and operations of the organisation more effectively? The broad aims of allocation procedures are to avoid under-utilisation or overcrowding. It means a constant redeployment of redundant spaces so as to serve new users.

Physical attributes such as location, size, shape and environmental factors define the use and thus encourage a more dynamic procedure for the allocation and reallocation of existing spaces over time (Musgrove and Doidge, 1970, p. 36). This becomes the means to meet changing business unit requirements within the organisation's premises, by the distribution of working groups within the building and the actual allocation of individuals to space. The allocation procedures take into consideration both the subdivision of space and design features involving stacking and adjacency.

The subdivision process in general inevitably deals with surplus space. Figure 10.2 shows the number of possible permutations in subdividing a very simple piece of space with a single core in the middle. The core represents the critical factor in floor access arrangements, and so should be configured to provide ease of access to the intended occupants.

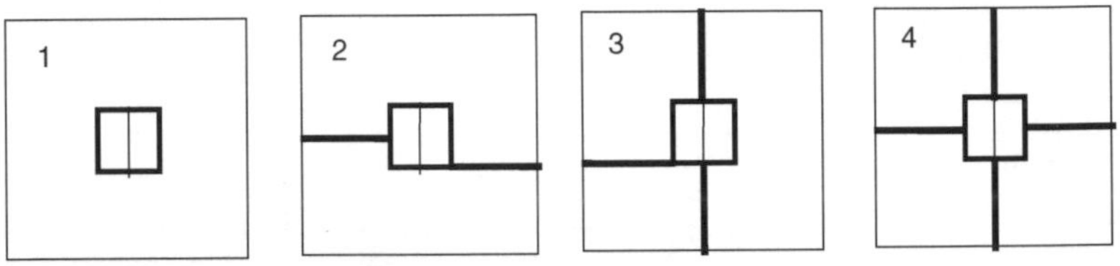

Source: Nutt (1993)

Figure 10.2 Subdivisioning, showing the possible levels of occupancy.

Complementary to the subdivision procedure is the profiling of a building by 'stacking' to accommodate the intended organisational functions, the departments or business units. This allows the distribution of departments and service infrastructures across floors, or the assignment of spaces to businesses. It permits the visualisation of each department in relationship to others (Binder, 1989, p. 10). When integrated with the adjacency concept, the synergy of working groups is enhanced. This could be implemented through compartmentalised areas or floors in the building.

But for a knowledge workplace, rather than aiming for the synergistic effects of adjacency planning, the notion of functional inconvenience may be adopted (Becker, 1990, p. 240). This concept has been developed through the inefficiencies of the adjacency pattern. Through such inconveniences, ideas and new products could be generated successfully within the organisation, as was shown already in the Steelcase CDC study (Becker, 1990, pp. 233–256). In summary, knowledge workplaces would probably see the reduction of dedicated allocation, or space assignments and potentially all spaces will veer to free address allocation as they cope with the knowledge exchange activities of the *k*-organisations. The knowledge workplace would also see to the complete provisions of space support.

10.5 CONCLUSIONS

We have now explored essential elements of knowledge work activities at the individual, team and organisational levels, justifying the view that guidelines to support learning are urgently needed in the workplace.

Since the nature of knowledge work is largely collaborative, space allocation will have to depart from the rank and status procedure that has been common practice in industry. In contrast, workplaces should provide flexible and intuitive support for multi-tasking work styles, whether the requirement is for privacy and concentrated work or for meeting people and socialising. Furthermore, each space would accommodate the necessary technology to ensure that 'place' of work is an information network that can be accessed from anyplace and anytime (Laing, 1996). The allocation procedure must be constructed from the types of knowledge exchange involving numbers of participants and the communications strategies to be adopted. Here, the needs of individuals, teams and organisation are differentiated at the outset.

On the other hand, when workplaces are already available, the consequences from the allocation and utilisation of the spaces will be the prime driver of workplace change. Ultimately, the allocation will be gauged from a period of evaluation to ensure that the allocated space is performing effectively and efficiently for the organisation, the end user, and management.

Our space allocation method is a fundamental approach. The continuum of allocation corresponds to variation in facility locations. It also matches the notion of availability over time—availability anytime and anyplace. Only then will the knowledge workplace matter to every level of the organisation. It is the task of the facilities management team to make use of organisational and facilities information as needed. Indeed, organisational issues such as attitudes, acceptance, and motivation are important. So are the physical ones: layout design, ICT, and synergy effects. But the prime goals should be how to support knowledge exchange effectively, and to ensure that employee learning is accelerated. The knowledge workplace is a provision for a diversity of options, and it is through training and education that users are prepared to adjust to the rules and constraint while exploiting the opportunities. We are just beginning to grasp some of the elusive points on what really matters in a knowledge workplace.

10.6 REFERENCES

Apgar, M., 1993, Uncovering your hidden occupancy costs, *Harvard Business Review,* May–June, pp. 124–136.

Apgar, M., 1998, The alternative workplace: Changing where and how people work, *Harvard Business Review,* May–June, pp. 121–138.

Association of College and Research Libraries, 1998, A Progress Report on Information Literacy: An Update on the American Library Association Presidential Committee on Information Literacy: Final Report. March. Available http://www.ala.org/acrl/nili/nili.html.

Avis, M. and Gibson, V., 1996, *Real Estate Resource Management Comparisons: A study of major occupiers in France, Spain, and the UK,* (Oxford: GTI Specialist Publishers).

Ballinger, R. and Gee, M., 1996, *Building knowledge in the workplace and beyond: A model National Workplace Literacy curriculum,* Revised edition, (Grayslake: Lake County College).

Bassi, L., Cheney, S. and Lewis, E., 1998, Trends in workplace learning: supply and demand in interesting times, *Training & Development,* **52**(11), pp. 51–74.

Becker, F., 1990, *The Total Workplace: Facilities Management and the Elastic Organization,* (New York: Van Nostrand Reinhold).

Becker, F. and Joroff, M., 1995, *Reinventing the Workplace,* (Atlanta: International Development Research Council).

Bertin, I. and Denbigh, A., 1996, *The teleworking handbook: new ways of working in the information society,* (Kenilworth: TCA).

Binder, S., 1989, *Corporate Facilities Planning: An Inside View for Designers and Managers,* (New York: McGraw–Hill).

Birchall, D. and Lyons, L., 1995, *Creating Tomorrow's Organization: Unlocking the Benefits of Future Work,* (London: Pitman).

Blackler, F., 1995, Knowledge, knowledge work and organizations: an overview and interpretation, *Organization Studies,* **16**(6), pp. 1021–1046.

Bradley, S. and Woodling, G., 1999, Accommodating future business intelligence: new work-space and work-time challenges for management and design, In *Conference Proceedings of Futures in Property and Facility Management, UCL, London, 24–25 June 1999,* 37–41.

Breivik, M., 1992, Information literacy: An agenda for lifelong learning, *AAHE Bulletin,* **44**(7), pp. 6–9.

Brown, B., 1998, Career development: A shared responsibility. ERIC Digest 201. Available http://www.ed.gov/databases/ERIC_Digests/ed423427.html.

Bruce, C., 1999, Information Literacy: An international review of programs and research. Paper for the Auckland '99, Lianza Conference, Nov 9–12. Available http://www.lianza.org.nz/conference99/bruce.htm.

Carnevale, A., 1989, The learning enterprise, *Training and Development Journal,* **43**(2), pp. 26–33.

Castells, M. and Aoyama, Y., 1994, Paths towards the informational society Employment structure in G-7 countries, 1920–90, *International Labour Review,* **133**(1), pp. 5–33.

Chadwick, A., 1993, Space/Time Office, *Facilities,* **11**(7), pp. 21–27.

Corcoran, M., 2001, But enough about me, what about the users? *Online,* **25**(6), pp. 90–92.

Davis, G., Becker, F., Duffy, F. and Sims, W., 1985, Executive Overview, An Overview of the Main Report and Supporting Volumes of the ORBIT-2 Project and rating Process on Organizations, Buildings and Information Technology, Harbinger Group Inc., Norwalk.

Digital, 1993, *Digital Annual Review* 1993, (London: Digital).

Duffy, F., 1992, *The Changing Workplace,* edited by Hannay, P., (London: Phaidon).

Eclipse Group, 1995, Workstyle 2000: Facilities Management Guide No. 9. In *Organisations of the Future*, edited by McLocklin, N., Maternaghan, M. and Lowe, S., (London: The Eclipse Group), pp. 5–10.

Egbu, C., Bates, M. and Botterill, K., 2000, A conceptual framework for studying knowledge management in project-based environments. In *Proceedings of the First International Conference of Postgraduate Research in the Built and Human Environment, 15–16 March 2001, Univ of Salford, UK*, pp. 186–195.

General Management Review, 2001, Evolution of a digital age organization, General Management Review, An Economic Times Presentation. Available http://www.etinvest.com/sm/gmr/kn_dig.htm.

Gordon, J., 1989, In search of … lifelong learning, *Training*, **26**(9), pp. 25–33.

Haron, L., 2000, Towards developing a facility space-time management method, *Journal of Valuation and Property Services*, **3**(1), pp. 1–22.

Howells, J., 2000, Understanding the new service economy. In *Knowledge and innovation in the New service economy*, edited by Andersen *et al.*, (Cheltenham: Edward Elgar), pp. 267–276.

Howells, J. and Roberts, J., 2000, Global knowledge systems in a service economy. In *Knowledge and innovation in the New service economy*, edited by Andersen *et al.*, (Cheltenham: Edward Elgar), pp. 248–266.

Imel, S., 1995, Workplace Literacy: Its Role in High Performance Organizations. ERIC Digest No. 158. Available http://www.ed.gov/databases/ERIC_Digests/ed383858.html.

Index to Coverage of Offices, 1964, *Architects' Journal*, 29 January.

Joroff, M., Louargand, M., Lambert, S. and Becker, F., 1993, *Strategic management of the fifth resource: Corporate real estate*, (Norcross: IDRF).

Laing, A., 1993, The Changing Workplace, The Flexible Workplace, *Proceedings of a one-day Seminar held at Digital Equipment Ltd., 13 July 1993*, edited by Yvonne, F., (Henley on Thames: Henley Management College), pp. 6–12.

Laing, A., 1996, Directions for change: the separation of office design for users, *Facilities Management*, **3**(6), pp. 18–19.

Langdon, F. and Keighley, E., 1964, User Research in Office Design, *Architects' Journal*, 5 February, 233 – 239.

Lankard, B., 1991, Worksite Training, ERIC Digest No. 109. Available http://www.ed.gov/databases/ERIC_Digests/ed329809.html.

Lloyd, B., 1993, The Future of Offices and Office Work. In *Responsible Workplace: The Redesign of Work and Office*, edited by Duffy, F., *et al.*, (London: Butterworth Estates Gazette), pp. 44–54.

McClure, C., 1997, Network literacy in an electronic society: An educational disconnect. In *Media literacy in the information age*, edited by Kubey, R., (New Brunswick: Transaction Publishers), pp. 403–439.

McDermott, R., 1999, Why information technology inspired but cannot deliver knowledge management, *California Management Review*, **41**(4), pp. 103–117.

Miles, R.E. and Snow, C.C., 1986, Organizations: new concepts for new forms, *California Management Review*, **18**(3), pp. 62–73.

Mosca, J., 1997, The re-structuring of jobs for the year 2000, *Public Personnel Management*, **26**(1), pp. 43–60.

Mukerji, D., 2000, *Managing Information: New Challenges & Perspectives*, Prentice Hall, Frenchs Forest.

Musgrove, J. and Doidge, C., 1970, Room Classification. Typewritten, UCL, London.

Nonaka, I. and Takeuchi, H., 1995, *The Knowledge-Creating Company,* Oxford University Press, Oxford.

Nutt, B., 1993, MSc. FEM, Unpublished Notes, Module 4.3.2, UCL, London.

Overtoom, C., 2000, Employability Skills: An update. Oct 1, 2000. ERIC Digest No. 220. Available http://www.ed.gov/databases/ERIC_Digests/ed445236.html.

Porter, M., 1999, Clusters and the new economics of competition. In *Managing in the new economy,* edited by Magretta, J., (Boston: Harvard Business Review Book), pp. 25–48.

Price, S., 1997, Facilities Planning: A perspective for the information Age, *IIIE Solutions,* **29**(8), pp. 20–22.

Quarstein, V., Ramakrishna, H. and Vijayaraman, B., 1994, Meeting the IT challenges of business. *Information Systems Management,* **11**(2), pp. 62–70.

Roberts, J., Miles, I., Hull, R., Howells, J. and Andersen, B., 2000, Introducing the new service economy. In *Knowledge and innovation in the New service economy,* edited by Andersen *et al.,* (Cheltenham: Edward Elgar), pp. 1–9.

Skagen, A. (ed.), 1986, *Workplace Literacy: AMA Management Briefing,* (New York: American Management Association).

Skyrme, D., 1994, Flexible Working: Building a Lean and Responsive Organization, *Long Range Planning,* **27**(5), pp. 98–110.

Spence, J., 1999, Worker-centered learning: Labor's role, ERIC Digest No. 211. Available http://www.ed.gov/databases/ERIC_Digests/ed434247.html.

Strassman, P., 1983, Information systems and literacy. In *Literacy for life: The demand for reading and writing,* edited by Bailey, R. and Fosheim, R., (New York: The Modern Language Association of America), pp. 115–121.

Stone, P. and Luchetti, R., 1985, Your office is where you are, *Harvard Business Review,* **64**(2), pp. 102–117.

Swap, W., Leonard, D., Shields, M. and Abrams, L., 2001, Using mentoring and storytelling to transfer knowledge in the workplace, *Journal of Management Information Systems,* **18**(1), pp. 95–114.

Thompson, J., 1997, The contingent workforce: the solution to the paradoxes of the new economy, *Strategy and Leadership,* **25**(6), pp. 44–45.

Varcoe, B., 1995, A Demanding Challenge, *Flexible Working,* 1(1), pp. 8–11.

Watts, R., 1994, Space Planning, *BIFM,* **1**(8), pp. 8–9.

Winslow, C. and Bramer, W., 1994, *Future work: Putting knowledge to work in the knowledge economy,* (New York: The Free Press, Maxwell Macmillan International).

Worthington, J., 1982, Using Office Premises Effectively. In *Business Property Handbook,* edited by The Boiscot Waters Cohen Partnership, (Aldershot: Gower).

Zeisel, J. and Maxwell, M., 1993, Programming Office Space: Adaptive Re-use of the H-E-B Arsenal Headquarters. In *Professional Practice In Facility Programming,* edited by Preiser, W., (New York: Van Nostrand Reinhold), pp. 153–181.

Zuboff, S., 1988, *In the age of the smart machine,* (New York: Basic Books).

CHAPTER 11

Towards Typologies of Knowledge Work and Workplaces

Reidar Gjersvik and Siri H. Blakstad

11.1 INTRODUCTION

During the last couple of decades, businesses that have been looking for new ways to enhance their performance have increasingly implemented new, innovate office solutions. Many of these businesses or organizations have been what we may term knowledge based service providers. Surprisingly, one may observe that very different organizations implement workplace solutions that are almost similar. Does this mean that they have the same requirements and needs, or is this a result of some workplace solutions being 'in fashion'? We argue that this is a result of a superficial understanding of knowledge work, and of how office space may support knowledge work.

The Knowledge Workplace (KWP) is a research initiative on new office solutions and new ways of working in knowledge intensive organizations. The goal of the Knowledge Workplace initiative is to develop knowledge about the relationships between: (i) organizing, organization development, and new ways of working; (ii) modern information and communication technologies, and (iii) architecture, new office solutions, and physical infrastructure; in knowledge based, knowledge intensive, and knowledge producing organizations and networks. We are particularly interested in studying change processes, both when it comes to the design and development of the solutions, and also the use of these solutions in value creating work.

KWP will consist of several projects in which organizations or networks develop, build, and use new workplaces and offices. Most of the research projects within the initiative will be action research projects (Greenwood and Levin, 1998; Skaret et al., 2001), in which the researchers act together with the organizational actors in producing change resulting in new work forms, new offices and new systems. Through this, new knowledge is co-generated (Elden and Levin, 1991). The cases we use to illustrate this paper come from such projects.

Through the research, we are aiming to develop a richer and more diverse vocabulary to describe and analyse knowledge work. We want to develop a better understanding of knowledge work in order to be able to provide better physical solutions that can support a range of knowledge work and a variety of organizations. In this paper we briefly present the use of archetypes in order to describe knowledge work, and elaborate on a typology of how knowledge work may be supported by space. In most cases, space is only one of several supporting mechanisms in the workplace. Information and communication systems, management and leadership, external services and other supportive functions will, together with space, define the environment in which knowledge work is performed. In the KWP, we deal with several of these issues, but here we shall concentrate on spatial support for knowledge work.

11.2 KNOWLEDGE INTENSIVE ORGANIZATIONS AND WORK

We focus our research on organizations that are knowledge based, intensive, and producing; we choose to use the term knowledge intensive organizations (Alvesson, 1995). Knowledge intensive organizations are commonly being distinguished by having knowledge as their primary input and output, and that the processes within the organization are mostly related to communication, coordination, processing of information, and the combination of knowledge. These organizations are often less related to heavy capital investments, like machinery; however, some of their work consists of the development and use of very sophisticated and expensive computers.

The products of these organizations are often hard to specify. Often the product itself is new knowledge, and since it is new, describing it is the same as creating it. The processes that produce such products are thus open ended. Instead of using a value chain approach to describe these, research shows that they may often be more profitably described as value networks or value workshops (Stabel and Fjellstad, 1998).

Alvesson (1995, p. 6) summarizes knowledge intensive organizations as being characterized by factors such as:

- Significant incidents of problem solving and non-standardized production;
- Creativity on the part of the practitioner and the organizational environment;
- Heavy reliance on individuals (and less dependence on capital) and a high degree of independence on the part of practitioners;
- High educational levels and a high degree of professionalism on the part of most employees;
- Traditional concrete (material) assets are not a central factor. The critical elements are in the minds of employees and in networks, customer relationships, manuals and systems for supplying services;
- Heavy dependence on the loyalty of key personnel and—this is the other side of the picture—considerable vulnerability when personnel leave the company.

Like Alvesson, we argue that these factors often are present in knowledge intensive organizations, but not all of them, and not always. Organizations that are close to routine work may also be knowledge intensive, because of the nature of the routines and the cases they are processing. Art or medical organizations are often also not considered in the common definition, but looking at what is being done, one cannot disregard the fact that the work is highly knowledge dependent.

Research done by Håkonsen and Carlsen (1999) within the KUNNE[1] research network indicates that knowledge work and knowledge intensive organizations are heterogeneous; they are as diverse as any kind of work or organization. It is not as though there exists one particular way of working that we may label 'knowledge work'. If we move away from the idea that there exists a certain kind of organization to be labelled 'the knowledge organization', we gain the freedom to explore various forms or types of knowledge work and the workplaces that may support these forms or types.

We have found that the models and language we have for describing work is often intended for orthodox organizations with processes that are closed, linear, and result in well-defined products. These ways of describing work may be suited for industrial or bureaucratic work processes, but are less relevant for knowledge work and knowledge intensive organizations as they have been described above.

11.3 IDENTIFYING KNOWLEDGE WORK ARCHETYPES

We suggest the use of knowledge work archetypes as a tool to describe various recognizable forms of knowledge work. To be more precise, what we are suggesting and demonstrating here is a structure for the creation of knowledge work archetypes, and how this may be related to a typology of knowledge workspaces.

The archetypes and the typology are intended as tools for describing individual organizations, so that the instances of archetypes and spaces that we present here do not claim any general relevance. Archetypes are commonly related to people; they are figures with a story, and they have certain personality traits that we recognize and identify with. Most people will identify with several archetypes, as each of them represents a particular quality or personality. Archetypes of knowledge work are, in a similar fashion, recognizable work patterns that we may identify in organizations. As with archetypes of people, organizations may be associated with several knowledge work archetypes. That is, one organization may have work patterns that include traits we associate with a number of different archetypes.

Archetypes are not definitions. Thus, we are not looking for distinctions by which we can divide knowledge work into a set of boxes that encompasses the total universe of knowledge work. Rather, we are looking for concepts that may be used to distinguish an archetype, to describe it as a recognizable type of work with a set of characteristics. This also implies that not all concepts will be used in order to distinguish all archetypes. The concepts may be considered building blocks that, together with examples and stories, are used to construct a knowledge work archetype that is recognizable. We also believe that the most powerful archetypes are the ones constructed through a process in the organization. The general archetypes we present here are more for illustrative purposes, as a starting point for a process in an organization, but not simply a menu from which to choose.

---

[1] KUNNE is a research network focusing on knowledge in organizations. It includes a number of research projects in close cooperation with more than twenty Norwegian companies and networks since 1997. To some degree, all these projects are action research projects, in which we learn with the companies as they go about making change processes. See http://www.kunne.no/english/index.htm.

We base our structure for the creation of knowledge work archetypes on an adaptation of a general model (see Figure 11.1) of knowledge based value creation from KUNNE (Skaret and Bygdås, 1999). The model illustrates that all organizations have a set of resources, and in the case of knowledge intensive organizations, we may label these (1) 'Knowledge resources'. The organization aims to produce various kinds of (2) 'Value and products' from these resources. The products, tangible goods or services, are what the organization produces for the customer. Value is not only economic value, it may be related to a diverse set of stakeholders, and may also be social, ethical, aesthetical or environmental value.

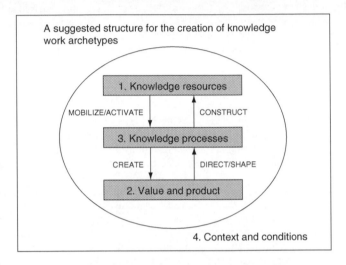

Figure 11.1 The structure of knowledge work archetypes.

In the KUNNE project, much of the focus has been on (3) 'Knowledge processes', through which resources are transformed into value. We have found that all too often, knowledge intensive organizations have been excessively focused on managing and monitoring their resources, instead of on the processes through which products and values are produced. And finally, as an adaptation of the Skaret and Bygdås model for our purposes, the interplay between resources, processes and values takes place within (4) 'Context and conditions' that are relevant for a given organization. The main elements of the model are to be found in Table 11.1.

Knowledge work archetypes are intended as tools that the user organizations can rely on for developing their workplace requirements. Next, we will describe a framework for defining a typology of knowledge workplaces. We suggest that work archetypes should be linked to a spatial typology that is based on what the user organization wishes that space should contribute to in the workplace.

Table 11.1 Main elements of the work archetype model.

Main element	Comprises
1. Knowledge resources	All the knowledge elements available to the organization, in terms of people, documents, routines, procedures, etc. The knowledge is carried in various forms; individual/collective, tacit/explicit (cf. Blackler, 1995).
2. Value and product	The different kinds of value that the organization creates for various external actors, e.g. customers, society. These joint objects or motives shape and give direction to the actors and the organization, both by their nature and by their characteristics, e.g. degree of innovation (Skaret and Bygdås, 1999).
3. Knowledge processes	Value creation takes place as knowledge resources are mobilized, activated and transformed. The processes have different characteristics: individual/collective work, internal/external orientation, connectivity, tools and systems, or speed/time/urgency (Wenger, 1998; Engeström, 1993; Blackler, 1995; Nonaka and Takeuchi, 1995).
4. Context and conditions	Knowledge work takes place in a relevant context, which both shapes and is shaped by the work. Certain conditions enable or prevent knowledge creation. Important aspects are community (practice, knowledge), organizational (structure, culture), technological (infrastructure, knowledge, systems), social (work life norms and values), and historical (events, stories) (Nonaka and Takeuchi, 1995; Nonaka and von Krogh, 2000; Wenger, 1998; Engeström, 1993).

11.4 A TYPOLOGY FOR WHAT SPACE DOES

During the last years, several typologies for classifying office buildings and space have been developed. Duffy (1997) proposes four types in a two-dimensional system, intending to link types of work and office space. He constructs four metaphors for different work-patterns: the Den, the Hive, the Cell and the Club. These are defined based on their degree of work autonomy and interaction: the Hive combines individual processes with a low level of both interaction and autonomy, the Den supports group processes and combines a high degree of interaction with a low degree of autonomy, the Cell is for concentrated study and matches a low level of interaction with a high level of autonomy, and finally the

Club represents transactional knowledge, showing a high level of both interaction and autonomy.

Another system for describing office space is Becker's (1999) terminology for alternative office solutions:

- Universal plan offices/workstations
- Activity-setting environments
- Non-territorial/unassigned offices/workstations
- Home-based telecommuting
- Team/collaborative environments
- Virtual officing.

A more precise way of describing office typologies has been developed at the University of Delft (Vos *et al.*, 1997). The Delft framework takes into account three dimensions: place, space, and use. Place (location) refers to central office or telework office (satellite office, business centre, guest office, home office, instant office); Space refers to cellular office, group office, open plan office or combined office, and Use is characterized as personal office, shared office (1:x) or non-territorial office (x:x). The three dimensions are an advantage of this framework; within the same physical solution, e.g. a cellular office, there can be different solutions for use (1:1 or shared) and several possible locations (central–decentralized).

Compared to the Becker and Delft schemes, our approach to typology will be less focused on how space is described physically (what space is), and more concerned with how it affects and contributes to the activities that take place within it (what space does). The reason for this is that we should be able to talk about what the user organization wishes to achieve by using workspace, before we start developing and designing physical solutions. In our experience, the range of possibilities will be narrowed if we discuss actual spatial solutions with the users before they have defined what they want to achieve. Options should be preserved at this stage. After discussing what space should do, it is time to start defining requirements according to how space functions. These requirements should be translated into spatial design alternatives, which the user organization can evaluate and develop further in co-operation with the designers.

There are several characteristics of space that may support knowledge work. So far, we have selected ten types of space functions that in our experience are important in knowledge intensive organizations.

11.4.1 Space for bringing people together

The main function of office buildings is to provide space for bringing people together. At the same time as the dependence of some activities on space and location is reduced by using technology to perform work from almost anywhere, the workplace becomes crucial as a place to meet and interact. When designing space, a few questions obviously need answers. Which functions rely on people meeting each other? Which activities are best performed from the workplace, and which are better performed in alternative locations? Do we need to be located together? How will people meet at the workplace—and where? Who will you need to meet—planned and unplanned?

In successful knowledge workplaces, we see a richer supply of informal and formal places to meet. Places to meet are often organized in a hierarchy: serving two to three people, workgroups, projects and teams, departments and ultimately the entire organization.

11.4.2 Space for production and sharing of knowledge

Physical conditions may impact on production and sharing of knowledge, as it can both encourage and hinder communication and contact. In order to enhance sharing and production of knowledge, one should encourage people to meet formally and informally, in one project and profession, across organizational and professional boundaries, and in a common workspace. Sharing of knowledge also relies on access to shared information.

11.4.3 Space for learning

Space for learning is in many ways related to space for production and sharing of knowledge, because learning relies on the sharing of knowledge and skills. In a work environment, learning is an exchange of knowledge between people and between functions and also training of employees, in particular those who are new. Exchange of 'tacit knowledge' may require space where people work together, because it is best passed on by 'doing' and observing how colleagues perform their work. This will require common space, possibilities to observe work done by others, as well as space to meet and discuss.

11.4.4 Space for communication

Space may encourage communication between people in the same room both formally and informally. But space may also impede communication, by separating people who need to see and talk to each other. Face-to-face communication is probably still the most important form, but communication tools such as videoconferences, digital communication, telephone, including conference calls, have changed the way people relate to and communicate with each other. In most cases, these new ways of communicating will require spatial support.

11.4.5 Space for co-ordination

In many knowledge intensive organizations, action relies on a common and shared knowledge process. One example is a production desk for a newspaper, where all the information has to be available to everyone, so that they can share decisions about the contents of next day's paper. Information and decisions must be shared and made available to all members of the group instantly, so that they all can act and develop their part of a unified solution. Such functions require real-time interaction and co-generated decisions that rely on information from several actors and according to which every member of the group has to act.

Space for co-ordination usually relies on a common workspace, including possibilities for individual processing and work within a common spatial setting. There will also be access to common and shared information systems, with information or plans often displayed on screens or walls in order to define a common source of information for discussion and decisions. An emphasis on visual and acoustic contact in order to encourage communication and interaction is often found. Space is often divided into project or task space.

11.4.6 Space for change

As organizations grow aware of how the workplace can be designed to match the work itself, they will discover how space can be changed along with organizational change. Knowledge work is in most cases dynamic, and many knowledge intensive organizations are confronted with markets and competitors that require them to be able to change quickly. Flexibility, the possibility of changing, has been one of the major drivers for innovative office solutions as we have seen them during the last decade. Most organizations want the possibility to change and adapt to new needs. Requirements for change and adaptations concern possibilities for growth and reduction in size, possibilities for arranging workspace for teams and projects, and accommodating expected changes in work practices. Requirements for more adaptable solutions can be met by changing the physical setting or building, by changing how the workplace is used, or by applying financial and contractual change strategies (Blakstad, 2001). To deal with change, a strategic way of programming and designing workplaces should be followed in order to assess the ability of alternative solutions to deal with future risk.

On the workplace level, we recognize the emerging needs for flexible, adaptable and adjustable solutions linked to standardization of equipment, less paper, shared electronic information systems and filing, the impact of an increased rate of people moving around, and work possibly performed in other locations.

11.4.7 Space for concentration

In Norway as in many other countries, the cellular office is perceived as the office archetype. For most employees, the cellular office is associated with privacy, concentration and status, all of which are seen as essential to their function in the organization. The cellular office provides space for individual concentration and confidentiality, but will at the same time separate the employees, resulting in less interaction and sharing of information. Individual privacy and concentration is set against shared learning, communication, and co-ordination.

New office solutions try to provide space for individual concentration in other ways than through traditional cellular offices. The main issues are those of providing space for focused individual work, shared or private, open or enclosed, with the appropriate degree of visual and acoustic privacy.

11.4.8 Space for confidentiality

Secrecy is a concern for managers and in general people who handle confidential information that should be kept from staff, customers or others. In workplace settings where openness, transparency, and access to shared information is the main design objective, it may be difficult to control the information that is seen or heard. Where confidentiality is an issue, the need for control must be recognized. Usually the need for confidentiality, if there is any, will be clearly expressed during the development of a knowledge work archetype. Design consequences will include physical and spatial support of confidentiality, by providing visual and acoustical privacy, and in relation to the handling of information, such a controlling the visibility of work on computer displays or printers. There will also be an issue of designing for 'safe' places to meet without being exposed to people outside the group.

11.4.9 Space for creativity

People can be creative in unusual places. This makes providing space for creativity a complex task. Creativity that requires several people to work together will rely on people meeting and interacting. Individual creativity may well be best stimulated away from office buildings and work desks. But some spatial contexts foster creativity more than others. Such stimulating environments will often provide safety, challenges and surprises, both esthetical and spatial. Design for workplaces that encourage creativity may include places to meet, formally and informally, places to be noisy and alternative places to work in a narrow traditional sense and also play, both socially and physically.

11.4.10 Space carrying meaning

In addition to the direct impact of space on knowledge work activities, space will always carry meaning. Seen in a social constructivist perspective, this meaning is related both to the physical workplace and to the process of making the workplace. Defining and designing the workplace constructs meaning. This will be materialized in the new workplace, and will influence future users and visitors, as well as those who are involved in making it. Workplaces can be designed in order to communicate meaning to users and to the public. The office will display how the organization appreciates its employees and what its core values are. Traces of activities will show up, signalling whether this is a vibrant and dynamic work environment or belongs to an organization frozen in practices of the past.

Identity and culture can be communicated by how the workplace displays itself to both employees and outsiders. This can be used consciously for marketing and for attracting employees and clients, and, intended or not, will signal how the organization relates to its employees and to society. Developing and using the corporate culture as a management tool has become prominent in knowledge intensive organizations. Buildings and space are strong communicators of such a culture.

The main asset of a knowledge intensive organization is the knowledge of its employees. When several organizations compete for the same skilled people, an attractive work environment may be useful to get hold of and to keep the people needed.

According to Argyris and Schön (1996), there are two ways of holding organizational knowledge: that organizations function as environments that capture, store, and sustain knowledge, and that they directly represent knowledge by theories of action. Organizations do function in several ways as holding environments for knowledge. This knowledge can be held in the minds of individual members, in files and records of policies, actions and regulations, and in the physical objects that the members use as references and guideposts as they go about their business. The physical workplace represents such physical objects, and the work environment may function as a holding environment for knowledge. There is information, or meaning embedded in an organizational environment that keeps traces of its past and present knowledge.

The process of defining space to match the needs of an organization will create meaning in two ways. The first is related to the workplace as a product of the construction process, and the other to the quality of the actual process. For the people occupying a space, the meaning and quality of that space is often just as dependent on how much they felt involved and appreciated during the process, as on the actual physical solution.

11.5 ARCHETYPES AND TYPOLOGY AS APPLIED TO THREE CASES

To illustrate the use of archetypes and typologies of space, we have chosen three cases from projects that we are currently working on. The cases are chosen to show the diversity of knowledge work. In particular, they illustrate how various forms of knowledge resources and processes constitute a recognizable form of work, and how these may be associated with spaces that do something for these archetypes. These cases are examples of action research, in which research methods are used within the change and design process, and also in order to reflect on and generalize the results.

The first two archetypes are associated with a project in a mutual insurance life company. This company is building new regional headquarters, and as a preparation for that, we have been involved in creating two office pilots for them to explore new ways of work. We use the method of archetypes in the participatory process of designing the new offices, both to communicate learning from the pilots and for the employees to describe their work. We construct the third archetype as a preparation for a new project in an oil company, which is going to add a new wing to an existing office building, creating space to receive the operating organization of a new oil and gas field. The process we will be involved in there is much the same as the one in the insurance company. During the process in both cases, we will match the archetypes with types of space to arrive at a suggested design together with the project architects and interior designers.

11.5.1 The Skilled Routines archetype

In the first case, there are insurance workers managing contracts and portfolios, as well as handling claims. Mainly, the work consists of repeating a limited number of tasks, but the mix of these tasks in each case combined with customer interaction calls for skilful application of experience based knowledge. The work is basically individual, shaped by ICT systems and case documents, and primarily driven by the work flow of the system, which again is driven by customers through the telephone, fax, e-mail, etc. In some cases, the work is directly customer driven, demanding everything from simple yes/no answers to heavy knowledge based consulting. Employees report that they depend on their colleagues for knowledge and experience transfer on new or difficult cases. This is particularly so in the training of new employees, which is based on an apprentice model. Table 11.2 gives an overview of the analysis and the design consequences.

Table 11.2 The Skilled Routines knowledge work archetype.

Aspect	Insurance claims processing and portfolio management
Knowledge resources	Embedded in systems, embodied in individual experience, encoded on paper.
Value and product	Economic results for company, risk reduction and financial safety for customers, collective knowledge for the company/department.
Knowledge processes	Mostly well known problems, individual processing, serial dependency, transfer of experiences and models of customer handling between individuals, customers may be in a difficult life situation, high work pressure and many new employees.
Context and conditions	Work shaped by systems and laws, all contact with customers through telephone, fax, and e-mail, documentation important.
Main objectives	Learning and communication were chosen as the most important objectives.
Space for sharing knowledge	Open plan, transparent.
Space for learning	Definition of workplace clusters, and possibilities of including newcomers in teams with more experienced workers.
Space for communication	In a common workspace. Places to meet, formally and informally, open and enclosed.

11.5.2 The Innovative Production Line archetype

Again, this case is taken from the insurance company, but here we are concerned with the development department, which develops new systems and products, mostly related to IT. The ideas for new products and systems come from either the customers or the organization, and they always need to be tested in the market. New developments must be thoroughly analysed, because, according to law, the company will have to maintain the systems for decades even if the product only sells for a short time; thus, they always risk a too diversified portfolio of systems. Work is mainly project based, meaning it is a task that is limited in time, with a more or less defined goal, and with a workgroup gathered for the purpose. The workgroup often includes employees from various departments, or external consultants. The work includes both extremely creative processes, like brainstorming, and extremely individualistic production, like programming. It resembles consulting, but all customers are internal. This archetype and its consequences are summarized in Table 11.3.

11.5.3 The Diverse Specialist Knowledge Team archetype

Here, we deal with the oil company and more exactly with its land-based operation of an offshore field. We find a team of well-educated specialists within different fields, having three main functions: to maintain the onshore side of drilling operations, to serve as knowledge based resource developers, and to act quickly in case of an emergency. The operations side consists of day-to-day preventive maintenance, in close cooperation with the offshore crew, including video meetings each morning. Resource development implies getting the maximum output from the oil field over a period of 10–20 years, in addition to doing analyses of the drilling and other geological information to ensure sustainability. This work has a reflexive research orientation over time. On the other hand, if anything happens to an oil well, they form an emergency team, which sits together in one room, and works on the problem continuously until it is solved and the well is in operation again. All these modes of work require very knowledgeable people, both highly educated (half of the staff hold PhD degrees) and highly experienced; you are required to work on the offshore drilling platform before you can work in onshore operations. Much of the work is IT intensive, using powerful workstations and several screens. See Table 11.4.

11.5.4 Other archetypes

The three archetypes described above are only examples of cases. In our work so far, we have recognized created a number of other work archetypes:

Flowing nomads (consultants, sales people). Flexible work, often travelling or working in customer offices.

Table 11.3 The Innovative Production Line knowledge work archetype.

Aspect	Insurance systems and product development
Knowledge resources	Embodied in individuals, embedded in development methods and existing technology, encultured in work patterns and internal network.
Value and product	New systems and products, mostly IT related. Internal customers only. Market, implementation, dependability, usability.
Knowledge processes	Development together with user/market departments, speed (time to market) is important, use of systems and methods. Individual 'hackers'.
Context and conditions	IT (consulting) community in an insurance environment, tight work market.
Main objectives	Facilitating projects was their most important objective. Project space, a common workspace, is given priority over the individual workspace. As projects change, people move. Most projects last 2–6 months, and a high level of flexibility is needed, because projects change continuously and also rely on creative processes and phases with more individual focus and concentration.
Space for bringing people together	Includes external consultants in teams, workplaces for guests in project rooms.
Space for learning, communication and co-ordination	Project room of different size and shape. Enclosed, shared team space with individual, temporary workplaces.
Space for change	Flexible size and furniture in order to rearrange project rooms as projects and teams change.
Space for concentration	Some cellular offices for people with special functions, 'library', quiet space for people with their workplace in project rooms.
Space for creativity	'Process rooms' for active, team-based work and discussions. Enclosed space equipped with the necessary technology.

Table 11.4 The Diverse Specialist Knowledge Team knowledge work archetype.

Aspect	Land-based operation of an offshore field
Knowledge resources	Embrained and embodied in specialists, encoded in data and models, encultured in work patterns and interpretative skills, embedded in strict quality procedures.
Value and product	Very high economic value, both short and long term. Knowledge assets, both internal and external. Safety and environment. Innovation for future exploration.
Knowledge processes	Highly complicated and diverse data field. Everyday routine maintenance (online with oil rig), drilling operations (high intensity team work), increased oil extraction, evaluation through interpretation and reflection, planning (large expertise teams), questions are asked and encouraged. Training/enculturing: on rig, in team.
Context and conditions	Competitive work market, highly dependent on human capital. Advanced technology (3D modelling). Engineer/professional community.
Main objectives	A plausible choice of objective is concentration. Within the chosen building layout, which is best utilized for open plan workplaces, this will be a process that balances the need for individual concentration against the physical possibilities in the building and the need for co-ordination, interaction, and learning.
Space for bringing people together	Across projects and professional boundaries. Formal and informal places to meet.
Space for co-ordination	Coordination of ordinary operations: daily meetings in large meeting rooms, equipped with voice and video communication, monitors, and project information. For important and dangerous operations (recurring, but of short duration): space for teams of varying size, with possibilities for analysing and simulating complex data, as well as for monitoring the ongoing offshore operation on powerful computers.
Space for concentration	Individual and enclosed? In addition to personal workplaces, there is a need for shared, extra powerful workstations.

Professional gurus with knowing support (physicians, performing artists). Work where one or a few key people own the work process, and often are the ones who meet the customer and are identified with the work. They depend on a skilled support team in order to perform their specialist work.

Integrity (medical/psychological/religious professionals). Professional work where trust and confidentiality is at the core of the performance.

Strategic and symbolic flow (top managers). A continuous series of actions that have strategic implications and that will have symbolic power in the organization and its environment. Often related to networking with partners, customers, and stakeholders. A mix between planned and emerging actions.

Outside-in knowledge network (consultants such as engineers, who work more with external partners than with colleagues). Professional organizations with employees who work in external projects with other professionals. Inside the organization, each employee appears to be working as an individualist, whereas they work externally in cross-functional teams.

Constructive team (marketing, research staff). Communication intensive teamwork, creating and producing new concepts and knowledge through processes of combining the existing with the new.

Incubator pizza anarchy (in science parks, 'greenhouses'). Small entrepreneurial organizations with common office facilities, often provided by the public or a group of investors. Business ideas, knowledge resources, and networks are broadly similar, and these organizations are expected to benefit from co-location. Often with young employees, long and strange working hours, and lack of routines.

11.6 CONCLUSIONS AND FURTHER RESEARCH

In our work we have seen that we need a more powerful descriptive language for use in the development and implementation of knowledge workplaces. In the KWP initiative, we have used knowledge work archetypes and a typology of workplaces as tools to aid communication when defining and designing knowledge workplaces. The knowledge work archetypes are recognizable forms of work. The typology of knowledge workplaces comprises types of spaces for archetypical work.

In our continued research, we expect to develop further the structure of the archetypes, also providing a list of archetypes of knowledge work, and types of space related to these. We also intend to refine the method or process of creating and presenting the archetypes of knowledge work and the types of space.

First, as we get more cases and more insight into knowledge intensive work, we need to refine and strengthen the model. In particular, we still do not have a good structure to represent and analyse knowledge processes, which is the core part of the model: the transformation of knowledge resources into products and value.

Second, the archetypes we have presented here are for illustration purposes. Whether there are universal archetypes with a more general validity remains to be investigated. We think it is neither fruitful nor possible to search for an exhaustive list of all knowledge work archetypes.

However, we do think that for purposes such as learning, benchmarking and process initiation, it will be useful to be able to list a set of archetypes of knowledge work, and relate these to types of space. This list could contain both archetypes that are commonly observed, and archetypes that represent extreme kinds of knowledge work.

Third, we believe that the most powerful archetypes are the ones constructed through a process in the organization in question. The creation of the local archetypes by the actors in the organization is a part of the process in which the local actors become aware of their own knowledge work, and attach meaning to the types of space created to facilitate the work. An issue for research, related to the use of our typologies in change processes, is how to make the model powerful, yet simple. How much to describe? Is it always fruitful to describe? Is it structure, or focus on the immediately recognizable?

As we have experienced in the project, defining archetypes is not enough to give directions to workplace design. In addition to an understanding of the work processes, organizational objectives are needed and a direction from which to prioritize various types of space functions. Defining these objectives should be where strategic management is strongly involved in the process.

Further research in KWP and KUNNE will focus on participation in defining, designing and implementing new workplaces for knowledge intensive organizations. We will continue both developing and evaluation the different cases. We are especially interested in developing further the process models, by discussing how we can implement the use of archetypes in a process, incorporating the different actors in the process and making them able to participate constructively.

11.7 REFERENCES

Alvesson, M., 1995, *Management of Knowledge-Intensive Companies,* (Berlin: de Gruyter).

Argyris, C. and Schön, D.A., 1996, *Organizational Learning: A Theory of Action Perspective,* (Reading, MA: Addison–Wesley).

Becker, F., 1999, Beyond alternative officing: Infrastructure on-demand. *Journal of Corporate Real Estate,* 1(2), pp. 154–168.

Blackler, F., 1995, Knowledge, Knowledge Work and Organizations: An Overview and Interpretation. *Organization Studies,* 16(6), pp. 1021–1046.

Blakstad, S.H., 2001, *A Strategic Approach to Adaptability in Office Buildings.* Doktor ingeniør thesis at the Norwegian University of Science and Technology, Trondheim, Norway.

Carlsen, A., 1999, Om kunnskap i KIFT-bedrifter. KUNNE nedtegnelser 03/99, SINTEF Industrial Management, Trondheim, Norway.

Duffy, F., 1997, *The new office,* (London: Conran Octopus).

Elden, M. and Levin, M., 1991, Co-generative Learning: Bringing Participation into Action Research. In *Participatory Action Research,* edited by Whyte, W.F., (Newbury Park, CA: Sage), pp. 127–142.

Engeström, Y., 1993, Developmental studies of work as a test bench of activity theory: The case of primary care medical practice. In *Understanding Practice: Perspectives on Activity and Context,* edited by Chaiklin, S. and Lave, J., (Cambridge: Cambridge University Press), pp. 64–104.

Greenwood, D.J. and Levin, M., 1998, *Introduction to Action Research: Social Research for Social Change,* (Thousand Oaks, CA: Sage).

Håkonsen, G. and Carlsen, A., 1999, Communities and Activity-Systems in Knowledge Intensive Firms. Paper presented at International Conference on Management of Information and Communication Technology, Scandinavian Academy of Management, Copenhagen, Denmark, September 15–16, 1999.

Nonaka, I. and Takeuchi, H., 1995, *The Knowledge–Creating Company,* (New York: Oxford University Press).

Nonaka, I. and von Krogh, G., 2000, *Enabling knowledge creation: How to unlock the mystery of tacit knowledge and release the power of innovation,* (Oxford: Oxford University Press).

Skaret, M. and Bygdås, A.L., 1999, Mobilizing Knowledge in a Knowledge Intensive Firm. Paper presented at Cistema–99: Mobilizing Knowledge in Technology Management, Copenhagen, Denmark, October 24–27, 1999.

Skaret, M., Sen, G. and Roberts, H., 2001, Diversity in Action Research. Paper presented at The Annual EGOS Conference, Lyon, France, July 5–7, 2001.

Stabell, C.B. and Fjeldstad, Ø., 1998, Configuring Value for Competitive Advantage: On Chains, Shops and Networks. *Strategic Management Journal,* **19**(5), pp. 413–437.

Vos, P.G.J.C., van Meel, J.J. and Dijcks, A.A.M., 1997, *The office, the whole office and nothing but the office: A framework of workplace concepts,* (Delft: Delft University of Technology, Department of Real Estate and Project Management).

Wenger, E., 1998, *Communities of Practice: Learning, Meaning and Identity,* (Cambridge: Cambridge University Press).

A European Workplace Knowledge Network

Keith Alexander

A.1 INTRODUCTION

The International Research Symposium enabled leading researchers to assess the extent to which current research contributes to advancing knowledge in Facilities Management.

Following the Symposium, members of the scientific committee met to assess the state-of-the-art of research in each of the thematic areas, to draw conclusions from the event, to agree an agenda for future work and consider opportunities for collaborative working.

The meeting coincided with the publication of the priority areas for the EC Framework 6 (FP6) programme of research and development. Discussion therefore focused on the contribution that Facilities Management research could make to the priority areas, opportunities for European funding and strengthening relationships in a 'network of excellence'.

Four key research areas of direct interest to Facilities Management were identified in FP6 framework:

1. eWork systems (1.1.2.i)—workplace design and management incorporating innovative technologies to facilitate creativity and collaboration;
2. Factory of the future (1.1.3.iii)—new processes and integrated production facilities;
3. Eco-buildings (1.1.6.i)—energy management and improve environmental quality as well as the quality of life for occupants;
4. Privatisation (1.1.7.ii)—managing municipal facilities through service partnerships.

The meeting resulted in proposals and an expression of interest, submitted to the EC, for the creation of a workplace knowledge network.

A.2 NETWORK OF EXCELLENCE

Current research effort in Facilities Management is fragmented, ad hoc and carried out in independent centres of excellence, each with distinct contributions to the field. Although some informal co-operation and collaboration has been achieved through the European Facilities Management Network, creation of a network of excellence would bring together the leading research centres, with complimentary skills and experience, to strengthen scientific and technological excellence and expertise. Formalisation of the research network would provide European leadership and create a world force in the field of Facilities Management.

The network would integrate Facilities Management research in Europe and seek to structure and shape the way that FM research is carried out in Europe, as an action-oriented, multi-disciplinary approach, with a focus on the design and management of the 'knowledge workplace', and concerns the interfaces amongst the organisation, work processes, information society technologies and the physical environment and support services.

The network would advance knowledge and develop web-based tools and systems for this emerging discipline. Although a focal theme for the network is the design and management of knowledge workplaces, Facilities Management considerations cut across discipline boundaries and identified priority areas in FP6. Facilities Management is applied across the private, public and voluntary sectors. Current applications span commercial, health, industrial and municipal facilities. Facilities Management research seeks a deeper understanding of key business issues in a knowledge economy—ownership, governance, flexible working, outsourcing and corporate social responsibility—and important socio-economic issues in a knowledge society—accessibility, best value, social inclusion and sustainability, in the context of the workplace.

Harnessing the knowledge and skills of the centres of excellence, will help realise the potential of Facilities Management in managing the impact of the structural and cultural change in organisations, associated with the two-fold transition toward a knowledge-based society and sustainable development. The 'knowledge workplace' nurtures creativity and innovation and contributes to improvements in the quality of work and working life, increases productivity and reduces risk.

A primary aim of the network would be to ensure the effective transfer of Facilities Management knowledge, skills and technology to SMEs and foster life long learning to promote improved access, understanding and usability of the built environment and support services.

Network management would co-ordinate activities including a collaborative programme of research projects, sector forums, research seminars and symposia. All partners would be linked to an interactive web-based platform for collaboration and knowledge management. The web platform would also provide opportunities for sharing of research methods and tools and exchange of research amongst centres, associate researchers and industrial partners. The network would provide training and support for researchers. The network would support the development of other centres of excellence, from amongst the associate partners and beyond.

1.3 PROPOSED RESEARCH AGENDA

1.3.1 Research programme

The proposed research programme will consider the combined impact of changing workplace factors – organisation, people, process, settings, technology and business services – to identify innovative ways of facilitating work.

The focus of the work would be on 'the creation and management of sustainable, collaborative and accessible (virtual and physical) settings for knowledge workers; and the development of integrated workplace strategies (which support diversity) for knowledge organisations across Europe.'

The work would be organised in four theme areas, aligned to priority areas in FP6—knowledge workplaces, integrated production facilities, sustainable workplaces and public service partnerships.

1.3.2 Knowledge workplaces

Knowledge workplaces—how do we create good workplaces for knowledge work taking into account organisational issues and relations, information and communication technology and the physical layout/design? Settings that enable interaction, promote creativity, etc.

Knowledge work, an essential activity in the new economy, driven and enabled by the Information and Communications Technology, has resulted in drastic changes in work organisation and in workplace provision. Various workplace arrangements, space support and utilisation strategies have accommodated these changes. But are these strategies relevant for knowledge workplace?

There are so many types of knowledge work, and so many different activities, that we need a more diversified vocabulary and solutions more based on the nature of work and organisation than on trends in office design and practice. We need to develop a better understanding of knowledge work in order to be able to provide better physical solutions that can support different kinds of knowledge work and different organisations.

The goal of the Knowledge Workplace initiative is to develop knowledge about the relationships between: organising, organisation development, and new ways of working; modern information and communication technologies; architecture, new office solutions, and physical infrastructure; in knowledge based, knowledge intensive and knowledge producing organisations and networks.

The network are particularly interested in knowledge about change processes, both when it comes to the design/development of the solutions, and the use of these solutions in value creating work.

Knowledge workplaces must support different modes of communication in knowledge exchange, inter and intra organisationally, at all operating levels through the full range of space allocation options, from *dedicated assignment* to *free address allocation*. This guidance should assist facilities management in planning for workplaces so as to fulfil diverse requirements among knowledge workers.

An environment in which human and artificial components provide a dynamic setting in which an adequate ICT design may be integrated to enhance performant learning: the clustering of uses in work environments using the workspace to foster the formation of services through networking. Here the new literacy challenges are supplemented with a method of exploring, through co-generative learning, the ambient readability of workplaces as a main resource to foster performant learning, through novel ICT solutions.

A.3.3 Integrated production facilities

The Workspace Project confirmed that workspace design and management, comes low on most enterprises' priorities. Therefore the topic does not appear as part of top-level decision-making, nor are there systematic decision or feedback processes to deal with workspace. More surprisingly, in the light of this, high quality, flexible and often cost-effective workspace is designed and built. This is due to the existence of a vast amount of experience and tacit knowledge at middle management level, but which is often brought in to action late and after other decisions have been made.

But the work disclosed many areas where workspace was clearly not optimal, as shown by the need for frequent adaptations, under-utilisation of space, space which obstructed the creation of integrated teamwork, and space which was not designed to embody more intangible, symbolic, values of importance to the workforce. This presented a challenge to existing ideas about the workspace. In particular, workspace satisfaction is not just about physical comfort and health and safety. Issues of participation, communication and control are at least as important.

Moreover it was clear that the tacit knowledge and experience would become less valuable as the pace of technical and market changes increased, and some of the more middle management personnel of industry retired. There would be fewer precedents and less access to 'tacit' knowledge; there would also be an increasing need for innovatory solutions that, without a more transparent and explicit decision/design process, would not be forthcoming.

For a number of the participating companies there were also important issues of long-term strategic space use and sustainable development. Many manufacturing sites are long-lived, and have evolved to meet the demands of changing markets and technologies. In recent years this evolutionary process has accelerated and issues of the adaptation of a legacy of buildings and other facilities to meet production for rapidly changing circumstances presents an important challenge.

As a response to changing markets many European manufacturers have, or are considering, outsourcing some, or all, of their manufacturing operations elsewhere in the world. The social and ethical implications of this are considerable. In terms of workplaces it means that long established sites that were often major local employers must either close or be adapted to new economic uses. This process of adaptation is another important challenge for the future.

These processes of adaptation can be developed to provide opportunities to nurture and support SMEs in industrial clusters, to ease the transition of the local economy and encourage enterprise in dependent communities.

A.3.4 Eco-buildings—the sustainable workplace

The network have a particular interest in application of the sustainable development agenda in the context of the workplace. The workplace is critical for sustainable development in that it brings together environmental, social and economic issues in ways that affect both future economic prosperity and the life-experience of individuals. Whilst we recognise the crucial role of technology in the development of the future workplace we also believe that the application of technology will not in itself lead to a sustainable future. Organisational, institutional, financial and social issues are of equal importance. Our approach is trans-disciplinary and cuts across many traditional boundaries. Methodologically we specialise in case study and action research, often carried out within the organisations of our members.

We have identified the following issues as significant areas for future research:

- Quality of life and productivity in the workplace. There are indications that although workplace productivity may be increasing, the quality of experience of the workplace may be deteriorating. The challenge is to develop workplace strategies that increase productivity whilst also increasing quality of life. This approach will integrate efficient energy and materials use studies with participative studies of users in the workplace.
- Whole-life issues. Although there is a growing body of knowledge concerning the performance of facilities through their lifecycles this has only partly been translated into effective strategies for their long-term management. This is an important area for work.
- Adaptive re-use of facilities. With the rise of global markets and the increasing pace of economic change existing facilities can rapidly become redundant. This is not just a problem for individual enterprises but also for the communities who may be economically dependent upon them. In this context effective strategies for the conversion of redundant specialist facilities to generic workspace are of great importance. Where existing facilities have cultural or heritage value this task is particularly challenging.
- Corporate social responsibility. Enterprises are increasingly recognising that they have a responsibility not only for sound financial management but also for their environmental and social impacts. The critical locus for these impacts is often the workplace. The development of methods of managing environmental and corporate risk in the context of the workplace is an important area for work.

A.3.5 Public service partnerships

As part of the FP6 programme of research and development, the EC are interested in a better understanding of the impact of a knowledge-based society on public service, citizens and governance.

Research and development is needed to support the development of forms of 'multi-level governance, which are accountable, legitimate, and sufficiently robust and flexible to address societal change including integration and enlargement, and to assure the effectiveness and legitimacy of policy making'.

Interest in the network focuses on public service facilities, and in their development as a community resource. The effect of different forms of privatisation, including outsourcing and contracting out, is of particular interest to some members of the network. However, it is important that the objectives of improving public service and providing best value predominate the appraisal of options.

Previous collaborative work in the network has sought to transfer knowledge and experience and to share best practice in outsourcing and contracting out in municipalities in Europe.

Further work is needed to develop appropriate FM 'business' models, and to identify the full range of options for service delivery partnerships. New forms organisation, should provide flexibility and accommodate change, propose the development of new services, offer full integration, and recognise the need for a smooth transition from more traditional public service models.

Research work in the network will address issues of public accountability (service and assets), articulation of responsibilities between service partners, and controls assurance (governance).

Index